Engineering Materials

This series provides topical information on innovative, structural and functional materials and composites with applications in optical, electrical, mechanical, civil, aeronautical, medical, bio- and nano-engineering. The individual volumes are complete, comprehensive monographs covering the structure, properties, manufacturing process and applications of these materials. This multidisciplinary series is devoted to professionals, students and all those interested in the latest developments in the Materials Science field, that look for a carefully selected collection of high quality review articles on their respective field of expertise.

Indexed at Compendex (2021) and Scopus (2022)

Raju Khan · Mohd. Abubakar Sadique ·
Shalu Yadav · Andrei Rotaru
Editors

2D Boron Nanosheets

Synthesis and Applications

Springer

Editors
Raju Khan
CSIR-Advanced Materials and Processes
Research Institute (AMPRI)
Bhopal, Madhya Pradesh, India

Mohd. Abubakar Sadique
CSIR-Advanced Materials and Processes
Research Institute (AMPRI)
Bhopal, Madhya Pradesh, India

Shalu Yadav
CSIR-Advanced Materials and Processes
Research Institute (AMPRI)
Bhopal, Madhya Pradesh, India

Andrei Rotaru
University of Craiova
Craiova, Romania

ISSN 1612-1317 ISSN 1868-1212 (electronic)
Engineering Materials
ISBN 978-981-97-7013-7 ISBN 978-981-97-7014-4 (eBook)
https://doi.org/10.1007/978-981-97-7014-4

This Springer imprint is published by the registered company Springer Nature Singapore Pte Ltd.
The registered company address is: 152 Beach Road, #21-01/04 Gateway East, Singapore 189721, Singapore

You have to grow from the inside out. None can teach you; none can make you spiritual. There is no other teacher but your own.

—Swami Vivekananda

Preface

The burgeoning field of two-dimensional (2D) nanomaterials has seen remarkable developments, with borophene emerging as a standout material due to its exceptional properties and potential applications. This book, *2D Boron Nanosheets: Synthesis and Applications*, provides a comprehensive exploration of borophene, its synthesis, properties, and a wide range of applications, drawing comparisons with other 2D nanomaterials and highlighting its unique advantages.

We begin with an in-depth analysis of the structural units of bulk boron crystals and 2D boron nanostructures, setting the stage for understanding the fundamental building blocks of borophene. This foundational knowledge is crucial as we delve into the advantages of 2D boron nanosheets over other 2D nanomaterials, where borophene's superior electronic, thermal, and mechanical properties are highlighted.

The chapter "Borophene: A New Analog of Graphene" underscores borophene's potential to revolutionize various industries, similar to graphene, but with unique attributes that open new avenues of research and application. The subsequent chapters on synthesis strategies provide a dual perspective, encompassing both computational/theoretical routes and experimental approaches, offering readers a holistic understanding of the methodologies employed to develop 2D boron nanostructures.

Properties of boron nanocrystals, borophene, and its polymorphs are meticulously detailed, presenting the distinct characteristics and variations within the material that influence its performance in different applications. The book then transitions to application-focused chapters, starting with ultrathin 2D boron nanosheets as energy materials, where we explore their potential in batteries, supercapacitors, and other energy storage devices.

Sensing applications of borophene are examined, highlighting its sensitivity and selectivity in detecting various chemical and biological agents. This is followed by a discussion on borophene as an emerging nanomaterial for biomedical applications, showcasing its promise in drug delivery, bioimaging, and other medical technologies.

The final chapter, "Potential Advancements and Upcoming Smart Research in Borophene: Challenges to Future," addresses borophene research's current limitations and future prospects. It provides a roadmap for overcoming existing challenges and propelling borophene towards becoming a cornerstone material in nanotechnology.

This comprehensive coverage aims to equip researchers, scientists, and students with a thorough understanding of borophene, its capabilities, and its future potential. We hope this book serves as a valuable resource for those looking to explore and expand the frontiers of 2D boron nanomaterials.

Bhopal, India Dr. Raju Khan
Bhopal, India Mr. Mohd. Abubakar Sadique
Bhopal, India Ms. Shalu Yadav
Craiova, Romania Dr. Andrei Rotaru

Acknowledgments

The completion of this book, *2D Boron Nanosheets: Synthesis and Applications*, has been a collective effort, and we are deeply grateful to all those who have directly or indirectly contributed to its development.

First and foremost, we would like to express humble thanks to the almighty for providing us with the necessary physical and mental strength, intellect, and understanding to carry out and accomplish the task. We extend our heartfelt thanks to the esteemed contributors whose expertise and dedication have enriched this book. Your meticulous research, innovative insights, and unwavering commitment to advancing the field of 2D boron nanomaterials have made this work possible.

We are profoundly grateful to our research institution and the head of the organization, Dr. Avanish Kumar Srivastava, Director, CSIR-Advanced Materials and Processes Research Institute, Bhopal, MP, India for his unwavering support. Your provision of resources, infrastructure, and an environment conducive to cutting-edge research has been instrumental in the realization of this project. We would like to express our sincere appreciation to our colleagues and peers for their constructive feedback, collaboration, and encouragement throughout the writing and editing process. Your intellectual generosity and shared passion for nanotechnology have been invaluable.

A special note of thanks goes to our families and friends for their patience, understanding, and support during the countless hours spent on this endeavor. Your belief in our work has been a source of inspiration and motivation. Raju Khan thanks his parents, wife Shazia M. Siddiqui, daughter Sara Khan, and son Aayan Khan for their everlasting love, enthusiasm for science, and encouragement to pursue every task successfully. Mohd Abubakar Sadique expresses heartfelt gratitude to God for his blessings and guidance throughout his life. He extends his deepest appreciation to his family members and friends. Ms. Shalu Yadav is thankful to the Almighty and acknowledges the support of her parents Mr. Sudhir Yadav and Mrs. Usha Devi, and her siblings Mr. Ashish Kumar and Ms. Bhavna Yadav, during the tenure of this work. The SRF fellowship offered by the CSIR is duly acknowledged by her.

We are also grateful to the editorial and production teams at Springer Nature for their professionalism and expertise in bringing this book to fruition. Your meticulous attention to detail and commitment to excellence have ensured that this book meets the highest standards.

Finally, we acknowledge the broader scientific community whose ongoing research and discoveries continue to inspire and challenge us. It is through collective efforts and shared knowledge that we can push the boundaries of science and technology.

Thank you to everyone who has contributed to this book. We hope it serves as a valuable resource and a catalyst for further advancements in the fascinating world of borophene and 2D boron nanomaterials.

Dr. Raju Khan
Mr. Mohd. Abubakar Sadique
Ms. Shalu Yadav
Dr. Andrei Rotaru

Contents

Editors and Contributors

About the Editors

Dr. Raju Khan currently working as a Senior Principal Scientist and Professor at CSIR-Advanced Materials Processes and Research Institute, Bhopal, Madhya Pradesh, India. He has a background in Chemistry and did his M.Sc. in Chemistry and Ph.D. in Chemistry from Jamia Millia Islamia, Central University, New Delhi, India. He has worked as a Post Doctorate at the University of the Western Cape, Cape Town, South Africa, and CSIR-National Physical Laboratory, New Delhi, India as a Young Scientist, Visiting Scientist at the University of Texas and Sanantonio, UTSA, USA. He has been awarded a Fellow of the Royal Society of Chemistry (FRSC) and also awarded a reputed BOYSCAST Fellowship, Government of India. He is the recipient of many prestigious achievements like being a Scientific Advisor for FSSAI, Government of India, a Life member of the Materials Research Society of India, and a Member of the International Advisory Committee, DST, Government of India. His work has also been highlighted in Nature India twice in 2008 for tracking food toxicity and recently in 2021 for the fabrication of electrochemical biosensors from carbon waste materials. He has published more than 180 articles with a high citation score of more than 5600 in several renowned journals including 40 book chapters in esteemed publishers, editing 25 books with various reputed publishers and patent file filled. He has several ongoing projects completed including international scientific collaborations with the USA, Czech Republic, and Russia.

Mr. Mohd. Abubakar Sadique is a research scholar at CSIR-AMPRI, Bhopal, India. His significant contributions to the SERB project "Development of Rapid Electrochemical-Based Diagnostics for the Detection of SARS-CoV-2 Infection" showcase his expertise in the urgent development of diagnostic tools during a global pandemic. Previously, he served as a Senior Project Fellow at CSIR-CEERI, Pilani, Rajasthan, India, where he worked on the mission-mode project "Design and Fabrication of Microfluidic-Based Bio-Sensors for Biochemical Detection," displaying his skills in the cutting-edge field of microfluidics. His research on nanotechnology in healthcare applications showcases his commitment to the scientific community. His extensive portfolio includes publications, research articles, and patents, transforming scientific research into practical applications and improving global health outcomes.

Ms. Shalu Yadav is currently pursuing her Doctorate from AcSIR-AMPRI, Bhopal under the guidance of Dr. Raju Khan. She has completed her post-graduation in M.Sc. (Chemistry) from Agra University, Agra, in the year 2017. She qualified CSIR-JRF and joined as a Ph.D. Chemical Sciences candidate in CSIR-AMPRI, Bhopal. Her research is primarily focused on the synthesis of two-dimensional (2D) nanomaterials and their application in the fabrication of biosensors. She has a keen interest in exploring how these advanced materials can be used to detect and monitor infectious diseases, a field that holds immense potential for public health advancements. She has authored several research articles in reputed journals, showcasing her findings and innovations. Additionally, she has edited one book and contributed to several book chapters, further highlighting her expertise and commitment to disseminating knowledge in her field.

Andrei Rotaru received the B.Sc. in Chemistry and Physics (2006) from the University of Bucharest (Romania) and subsequently obtained the Doctorate of Engineering in Mechanical Engineering (2011) from the University POLITEHNICA of Bucharest (Romania), the Ph.D. in Chemistry (2013) from University of St. Andrews (UK) and the Doctorate of Science in Physics (2016) from the University of Craiova (Romania). Between April 2014 and October 2015, he was a Postdoctoral Researcher at the University of Craiova with a fellowship as Visiting Researcher at the University of Cambridge (UK). In 2006 he joined INFLPR-The National Institute for Laser, Plasma and Radiation Physics (Romania) as an Assistant Researcher and was promoted to Senior Researcher II in 2014. Further, in 2017 he became an Associate Professor of Chemistry at the University of Craiova, and since 2019 he has also been a Senior Researcher II at the Institute of the Physical Chemistry Ilie Murgulescu of the Romanian Academy. In 2021 he defended the Habilitation at the University of Craiova. Currently, he is the President of ESTAC (European Society for Thermal Analysis and Calorimetry), President of CEEC-TAC (Central and Eastern European Committee for Thermal Analysis and Calorimetry), and Chairman of the Scientific Commission of ICTAC (International Confederation of Thermal Analysis and Calorimetry). He is the Editor-in-Chief of Hybrid Advances and Coordinating Editor of Applied Surface Science Advances, serving also as a Member of the Editorial Boards of the journals: Surfaces and Interfaces, Journal of Thermal Analysis and Calorimetry, and Molecules. His research interests include Solid state chemistry, Condensed matter physics, Thermophysical properties of materials, Dynamics of nonlinear systems, Surface Science, and Processing thin films of soft materials by laser-assisted techniques.

Contributors

Usama Ansari Department of Chemistry, Netaji Subhas University of Technology (N.S.U.T), Dwarka, New Delhi, India

Lal Singh Banjara Govt Indira Gandhi Home Science Girls PG College, Shahdol, India

Derajram Benival National Institute of Pharmaceutical Education and Research (NIPER), Ahmedabad, India

Jutishna Bora Amity Institute of Biotechnology, Amity University, Jharkhand, Ranchi, India

Nimeet Desai National Institute of Pharmaceutical Education and Research (NIPER), Ahmedabad, India

Anshika Gupta Biomaterials and Sensor Laboratory, Department of Physics, CCS University Meerut Campus, Meerut, Uttar Pradesh, India

Shivam Kashyap Department of Chemistry, Netaji Subhas University of Technology (N.S.U.T), Dwarka, New Delhi, India

Raju Khan Industrial Waste Utilization, Nano and Biomaterials, CSIR—Advanced Materials and Processes Research Institute (AMPRI), Bhopal, Madhya Pradesh, India;
Academy of Scientific and Innovative Research (AcSIR), Ghaziabad, India

Shubhangi Khode National Institute of Pharmaceutical Education and Research (NIPER), Ahmedabad, India

Nagavendra Kommineni Center for Biomedical Research, Population Council, New York, NY, USA

Deepak Kumar Biomaterials and Sensor Laboratory, Department of Physics, CCS University Meerut Campus, Meerut, Uttar Pradesh, India

Shivam Kumar Amity Institute of Biotechnology, Amity University, Jharkhand, Ranchi, India

Sumit Kumar Department of Chemistry, Netaji Subhas University of Technology (N.S.U.T), Dwarka, New Delhi, India

Smita Lata Amity Institute of Biotechnology, Amity University, Jharkhand, Ranchi, India

Sumira Malik Amity Institute of Biotechnology, Amity University, Jharkhand, Ranchi, India

Sagar Mondal Amity Institute of Biotechnology, Amity University, Jharkhand, Ranchi, India

Pashupati Pratap Neelratan Biomaterials and Sensor Laboratory, Department of Physics, CCS University Meerut Campus, Meerut, Uttar Pradesh, India

Shreya Pande Department of Biomedical Engineering, Indian Institute of Technology Hyderabad, Kandi, India

Pranav Patel National Institute of Pharmaceutical Education and Research (NIPER), Ahmedabad, India

Deepak Poddar Department of Chemistry, Netaji Subhas University of Technology (N.S.U.T), Dwarka, New Delhi, India

Swati Priya Amity Institute of Biotechnology, Amity University, Jharkhand, Ranchi, India

Dhwani Rana National Institute of Pharmaceutical Education and Research (NIPER), Ahmedabad, India

Himanshu Ranjan Amity Institute of Biotechnology, Amity University, Jharkhand, Ranchi, India

Pushpesh Ranjan CSIR—Advanced Materials and Processes Research Institute (AMPRI), Bhopal, India;
Academy of Scientific and Innovative Research (AcSIR), Ghaziabad, India

Sarvesh Rustagi Department of Food Technology, Uttaranchal University, Dehradun, Uttarakhand, India

Mohd. Abubakar Sadique Industrial Waste Utilization, Nano and Biomaterials, CSIR—Advanced Materials and Processes Research Institute (AMPRI), Bhopal, MP, India;
Academy of Scientific and Innovative Research (AcSIR), Ghaziabad, India

Sagar Salave National Institute of Pharmaceutical Education and Research (NIPER), Ahmedabad, India

Siddharth Salve National Institute of Pharmaceutical Education and Research (NIPER), Ahmedabad, India

Anjana Sarkar Department of Chemistry, Netaji Subhas University of Technology (N.S.U.T), Dwarka, New Delhi, India

N. Sathish CSIR - Advanced Materials and Processes Research Institute (AMPRI), Bhopal, India;
Academy of Scientific and Innovative Research (AcSIR), Ghaziabad, India

Sanjeev K. Sharma Biomaterials and Sensor Laboratory, Department of Physics, CCS University Meerut Campus, Meerut, Uttar Pradesh, India

Monu Kumar Shukla Drug Chemistry Research Centre, Indore, Madhya Pradesh, India

Ankita Singh Department of Chemistry, Netaji Subhas University of Technology (N.S.U.T), Dwarka, New Delhi, India

Archana Singh CSIR—Advanced Materials and Processes Research Institute (AMPRI), Bhopal, India;
Academy of Scientific and Innovative Research (AcSIR), Ghaziabad, India

Sanjeeve Thakur Department of Chemistry, Netaji Subhas University of Technology (N.S.U.T), Dwarka, New Delhi, India

Harshita Tiwari Drug Chemistry Research Centre, Indore, Madhya Pradesh, India

Rudramani Tiwari Department of Chemistry, Institute of Science, Banaras Hindu University, Varanasi, India;
Regional Ayurveda Research Institute, Amkho, Gwalior, India

Shalu Yadav Industrial Waste Utilization, Nano and Biomaterials, CSIR—Advanced Materials and Processes Research Institute (AMPRI), Bhopal, MP, India;
Academy of Scientific and Innovative Research (AcSIR), Ghaziabad, India

Structural Units of Bulk Boron Crystal and 2D Boron Nanostructures

Shalu Yadav, Mohd. Abubakar Sadique, Pushpesh Ranjan, N. Sathish, and Raju Khan

Abstract Boron is an intriguing element that can take many different shapes, from bulk crystals to two-dimensional (2D) nanostructures. Furthermore, boron is the element that is next to carbon on the periodic table. It has the unique ability to form strong covalent two-center-two-electron bonds (2c-2e$^-$) and stable electron-deficient multi-center-two-electronc bonds, which means that it can form many unique 2D structures that are different from other 2D materials that have been studied. Recently, 2D boron crystals have been the object of intense interest due to the successful synthesis of atomically thin crystalline 2D boron sheets (i.e., borophene). Understanding the arrangement of boron atoms in these different structures is essential for unlocking their potential in various applications, such as electronics and energy storage. By exploring the structural units of bulk boron crystals and 2D boron nanostructures, we can gain insight into their properties and potential uses in cutting-edge technologies.

Keywords Boron · Borophene · Allotropes · Boron crystals · Two-dimensional nanostructures

1 Introduction

After successfully synthesizing graphene, numerous research efforts have been initiated to investigate the growth, structure, and properties of various two-dimensional (2D) materials. 2D materials possess exceptional physical and chemical characteristics that are utilized in nanoelectronics, optoelectronics, energy storage, and thermoelectric and thermal control [1]. Boron, the fifth element on the periodic table, possesses both semiconductor and metalloid characteristics. It is located in Group III A, adjacent to carbon, and shares a comparable electrical structure with carbon

S. Yadav · Mohd. A. Sadique · P. Ranjan · N. Sathish · R. Khan (✉)
CSIR - Advanced Materials and Processes Research Institute (AMPRI), Bhopal 462026, India
e-mail: khan.raju@gmail.com; khan.raju@ampri.res.in

Academy of Scientific and Innovative Research (AcSIR), Ghaziabad 201002, India

R. Khan et al. (eds.), *2D Boron Nanosheets*, Engineering Materials,
https://doi.org/10.1007/978-981-97-7014-4_1

atoms [2]. Due to its proximity to carbon in the periodic table, boron is anticipated to have independent monolayer allotropes. The chemical characteristics and complex structure of boron make it a desirable material. There are presently five known forms of boron, and its ground state structure has been a source of debate up to this point [3]. Nevertheless, the instability of a boron honeycomb sheet, like graphene, arises from the presence of four orbitals in its trivalent outer shell, which hampers the establishment of a closed-shell electronic configuration [4]. While boron exists in sixteen known forms, pure boron allotropes have been found for the β-rhombohedral, α-rhombohedral, two tetragonal, and the newly found γ-rhombohedral phase under high pressure. Additionally, numerous forms of boron, including bulk boron, clusters, nanotubes, monolayers, quasi-planar sheets, and bilayer sheets, have been the subject of both theoretical and experimental research [5]. Scientists have recently succeeded in producing unique boron nanowires and nanobelts. These materials exhibit properties indicative of semimetal semiconductors or narrow-gap semiconductors, according to the experimental results. The exact atomic structures of these materials are still unknown, though. The review of the literature reveals that not much work has been done on the planar geometry of boron crystals, specifically those in which B_{12} icosahedra are substituted [6].

The investigation of boron in various compounds has been ongoing for centuries as a result of its extraordinary capacity to form compounds with nearly all other elements. Hexagonal boron nitride (h-BN) is a III-V chemical with a large bandgap. The structure of h-BN is reminiscent of graphite, with alternating layers of nitrogen and boron atoms. h-BN exhibits superior thermal conductivity, exceptional physical characteristics, and extraordinary chemical stability [7]. It is reasonable to anticipate the production of purified boron, given its close resemblance to graphite. Sands' work in 1957 was the first to establish the existence of purified boron. Furthermore, Sands disclosed the identification of bulk g-B106, which boasts of an exceptionally intricate structure. Nevertheless, the crystal structures of only a small number of these polymorphs are known [8].

The chapter explores the fundamental aspects and advancements of boron in both bulk and 2D forms. It delves into the crystal structure, bonding, and classification of various 2D boron structures, highlighting their structural diversity and versatility. The chapter also discusses synthesis methods for 2D boron nanostructures, presenting theoretical and experimental approaches. It also discusses the topic of tailoring properties of 2D boron nanostructures, focusing on electronic, mechanical, and chemical properties for specific applications. The chapter concludes by discussing the applications of bulk boron and 2D boron nanostructures, aiming to inspire further research and innovation in the field. Figure 1 shows the schematic representation of the synthesis methods, modification, and applications of 2D boron nanostructures [9].

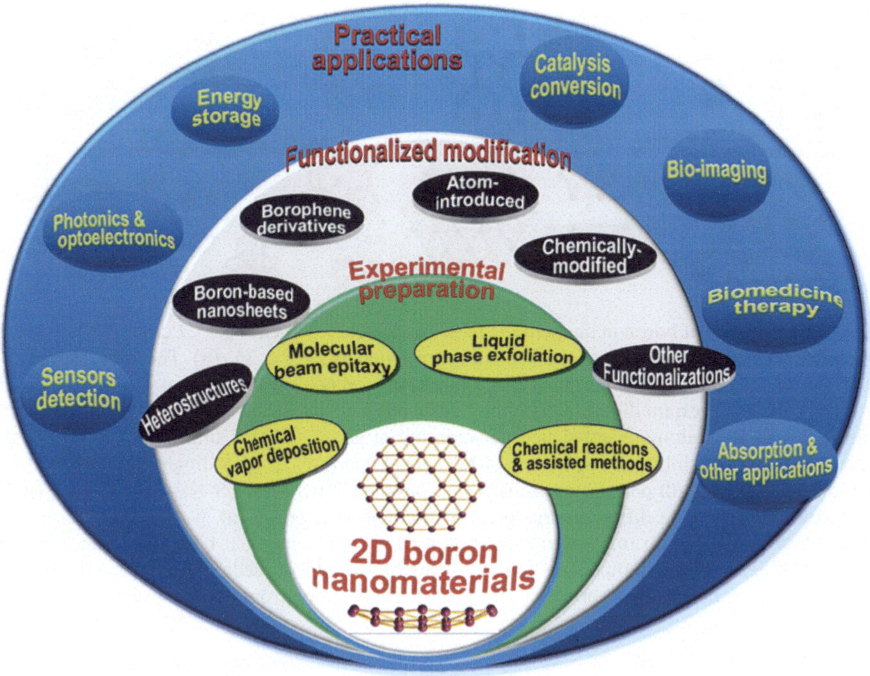

Fig. 1 The graphical representation of synthesis methods, derivatives, and various applications of 2D boron nanomaterials. Reprinted (adapted) with permission from [9]. Copyright 2022 Elsevier

2 Crystal Structure of Bulk Boron

2.1 Elemental Boron Allotropes

Boron stands out as a distinctive element in the Periodic Table due to its diverse bonding characteristics in various allotropes. The B_{12} icosahedron, rather than the individual boron atoms seen in metals, is the fundamental unit of elemental boron. The occurrence of gaps in the solid boron structure due to inadequate icosahedra packing leads to its low density. Elemental boron can react with a variety of non-metallic elements, forming binary compounds with several practical applications. For example, boron carbide (B_4C) plates can withstand a 30-caliber, armor-piercing bullet while weighing 10–30% less than typical armor [10]. Additional significant boron-nonmetal compounds include boron nitride (BN), which is formed by reacting boron with an excess of nitrogen, boron oxide (B_2O_3), which is produced by heating boron with an excess of oxygen, and boron trihalides (BX_3), which are formed by heating boron with an excess of halogen [11]. There are several different forms of elemental boron, and they all have structures with icosahedral subunits. The icosahedral components of these formations have partially filled interstitial spaces between

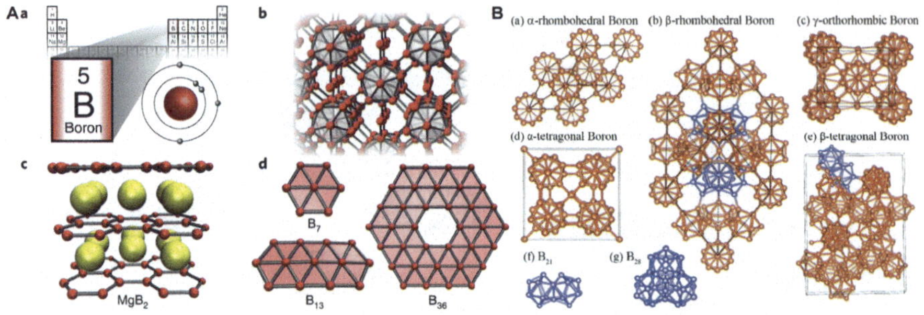

Fig. 2 The location of boron in the periodic table of elements is in Group 13, Period 2. The orbital structure of boron's trivalent outer shell can be represented as $1s^2\ 2s^2\ 2p^1$ (**a**). The atomic arrangement models consist of bulk boron in the γ28-phase, comprising icosahedral B_{12} clusters and B_2 dimers (**b**); magnesium diboride exhibiting graphene-like atomic arrangements having magnesium layers (yellow atoms) sandwiched between boron atoms (**c**); and diminutive boron quasi-planar clusters that exhibit the emergence of hexagonal apertures as their dimensions expand (**d**). Reprinted (adapted) with permission from [13]. Copyright 2018 Springer Nature, (B) The boron allotropes consist of five different structures: α-rhombohedral boron (**a**), β-rhombohedral boron (**b**), γ-orthorhombic boron (**c**), α-tetragonal boron (**d**), and β-tetragonal boron (**e**). The diagram also depicts the structural components of the B_{21} double-fused icosahedron (**f**) and the B_{28} triply-fused icosahedron boron (**g**). Reprinted (adapted) with permission from [14]. Copyright 2017 John Wiley & Sons

them. The number of atoms in the unit cell varies according to the use of these interstitial gaps [12]. Figure 2a shows the location of boron in the periodic table of elements in Group 13, Period 2. The orbital structure of boron's trivalent outer shell can be represented as $1s^2\ 2s^2\ 2p^1$. In addition, atomic structure models include bulk boron in the γ28-phase, which is made up of icosahedral B_{12} clusters and B_2 dimers and the magnesium diboride, consisting of graphene-like honeycomb networks of boron sandwiched between close-packed magnesium layers (yellow atoms) and small boron quasi-planar clusters that develop hexagonal holes as their size increases [13]. Moreover, the different allotropes for borophene and the interaction that existed between their boron atoms are also illustrated in Fig. 2b [14].

2.2 *Crystallography of α-Boron (Most Stable Form)*

α-Boron, also known as rhombohedral boron, is the most stable crystalline form of boron at ambient pressure and temperature conditions. It has a trigonal crystal system and an R-3 m (Hermann-Mauguin notation) space group. The unit cell parameters are; a $\approx$ 506 pm, b $\approx$ 506 pm, c $\approx$ 506 pm, and $\alpha \approx \beta \approx \gamma \approx 58.06°$. There are two inequivalent boron sites (B1 and B2) within the unit cell:

- **B_1:** 6-coordinate, surrounded by six other boron atoms at varying bond distances (1.67–1.80 Å)

- **B_2:** 7-coordinate, surrounded by seven other boron atoms with two shorter bonds (1.77 Å) and two longer bonds (2.01 Å)

α-Boron possesses a complex three-dimensional network structure formed by interconnected icosahedral boron clusters. These clusters are slightly distorted and share faces and edges with neighboring clusters, forming a continuous network [15]. The unique attributes of α-Boron include:

- **High Packing Density:** The arrangement of boron atoms in α-Boron leads to a high packing density, making it one of the hardest elements known. Even at ambient temperature, the crystal structure of α-boron remains consistent throughout the entire pressure range that has been tested. There have been no observed phase transitions, including the anticipated transition to α-Ga type. The structural characteristics demonstrate a consistent fluctuation in response to pressure fluctuations [16]. Li et al. predicted a high-pressure, close-packed boron phase using ab initio evolutionary techniques. Phase B_{10} has an hexagonal P63/mcm symmetry. B_{10} is more stable than α-Ga above 375 GPa and maintains consistency for at least 500 GPa [17].
- **Semiconductor Properties:** Despite being a network solid, α-Boron exhibits semiconducting behavior due to the presence of overlapping p-orbitals on the boron atoms. In terms of total energy calculations, α*-boron is less stable than α-boron but more advantageous than β-boron and γ-boron at zero pressure. We have confirmed that both α-boron and α*-boron are mechanically and dynamically stable. The mechanical and electrical properties of α-boron and α*-boron indicate that they have the potential to be superhard semiconducting phases of the element boron [18].
- **Anisotropy:** α-Boron exhibits anisotropic properties, meaning its physical characteristics (electrical conductivity, thermal expansion, etc.) vary depending on the direction within the crystal. As anticipated, the thermal expansion of α-rhombohedral boron and isostructural boron-rich pnictides was discovered to be anisotropic in several crystallographic directions [19].

Furthermore, another study examines how free energy fluctuates in response to pressure. In comparison to β-boron, α-boron demonstrates a higher degree of stability at absolute zero. This remains unaltered, even when zero-point energy and atomic disorders are considered [20].

2.3 *Bonding in Bulk Boron*

Bulk boron, specifically the most stable form α-Boron, exhibits a unique and complex bonding scheme. Unlike many elements that form well-defined covalent or ionic bonds, α-Boron lacks classical localized bonds between its atoms. Instead, it displays a type of bonding called multicenter bonding. Multicenter bonding involves the delocalization of electrons across multiple boron atoms within the crystal structure. The p-orbitals of neighboring boron atoms overlap extensively, leading to a "sea" of delocalized electrons. This delocalization allows for a more efficient filling of available bonding orbitals, contributing to the stability of the crystal structure [21]. The α-Boron structure consists of interconnected icosahedral clusters of boron atoms. Within these clusters, boron atoms exhibit a combination of covalent and metallic bonding characteristics. The delocalized electrons contribute to a metallic conductivity, despite being a network solid. Due to the directional nature of the p-orbitals involved in bonding, α-Boron exhibits anisotropic properties. This means its physical characteristics (electrical conductivity, thermal expansion, etc.) vary depending on the direction within the crystal. Unlike simple covalent solids like diamond (C–C single bonds), α-Boron lacks a well-defined Lewis structure. It also differs from ionic solids like sodium chloride (Na^+ Cl^- electrostatic attraction) where there's a clear distinction between cations and anions. Since Boron has multicenter bonding, there are numerous consequences. The delocalized electron network and interconnected clusters contribute to the extreme hardness of α-Boron. While the delocalized electrons provide some metallic character, the band gap is large enough to classify α-Boron as a semiconductor [22].

The bulk allotropes of boron display unique bonding characteristics as a result of their electron shortage. In the case of the intra-icosahedron boron, each boron atom possesses only three valance electrons. However, it can make bonds with five surrounding atoms by creating three-center-two-electron bonds (3c-2e^-). These bonds have an electron density maximum positioned near the center of the boron triangle. In α-rhombohedral boron, the icosahedra are linked together by both 3c-2e^- and 2c-2e^- bonds. In the case of γ-orthorhombic boron, the icosahedra are connected by 2c-1e^- bonds, whereas strong 2c-2e^- and 3c-2e^- connections exist inside the B_2 pair and between B_2 and the icosahedra. Furthermore, it has been demonstrated that boron allotropes can exhibit ionicity. The ionic nature of γ-orthorhombic boron was first described by Oganov et al. In this configuration, the B_{12} icosahedron acquires electrons while the B_2 pair loses electrons, similar to a rock salt. Subsequent research, however, has confirmed that the predominant form of bonding in γ-orthorhombic boron is covalent [14].

In recent years, size-selected boron clusters have received attention for their fascinating topologies, bonding characteristics, and potential usage as nanomaterial motifs. Using boron cluster anions (Bn −) has enabled extensive research using photoelectron spectroscopy and theoretical analysis, revealing mostly two-dimensional structures. The planar B_{36} cluster with a core hexagonal vacancy was the first empirical proof of two-dimensional borons, introducing the term borophene [23]. Chemical bonding models for boron clusters reveal stability and metal-centered boron wheels with 10 planar coordination numbers. These molecular wheels reveal boron-transition metal interactions and boron transmutation, increasing the potential for discovering new structures and materials through B^-/C comparison [24]. The typical chemical bonding that exists in the B_{36} polymorph of borophene is depicted in Fig. 3b [25].

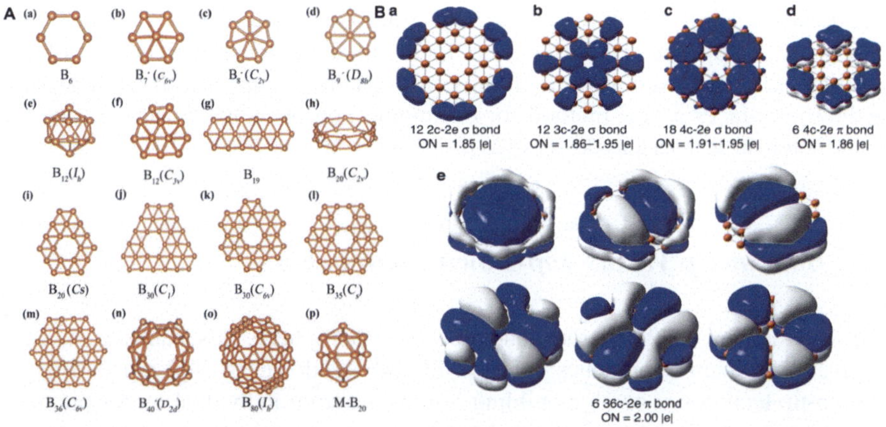

Fig. 3 **a** Some typical structures and the bonding of boron clusters. Reprinted (adapted) with permission from [14]. Copyright 2017 John Wiley & Sons. **b** Analyzing the chemical bonding in the hexagonal C_{6v} structure of B_{36}. The analysis was conducted utilizing the AdNDP methodology. The ON represents the occupancy number, precisely 2.00 |e|, indicating a density of exactly two electrons in the ideal scenario. Reprinted (adapted) with permission from [25]. Copyright 2014 Springer Nature

3 Diverse World of 2D Boron Nanostructures

In addition to three-dimensional (3D) boron icosahedral solids, 2D boron also displays numerous distinct structures as it is electron deficit, distinguishing it from other widely recognized 2D materials. In general, 2D boron crystals can be categorized into three distinct groups: (1) Atomically monolayered boron sheets resembling graphene, (2) 2D boron structures consisting of only one or a few unit cells in thickness, and (3) recently discovered novel 2D boron structures [14]. Furthermore, Tai et al. documented the production of thin 2D icosahedral boron sheets. The production of large-area 2D boron crystals has generated significant enthusiasm within the fields of boron chemistry and 2D materials [26].

4 Synthesis Methods for 2D Boron Nanostructures

Researchers have developed a new method for manipulating characteristics in 2D materials by designing atomic structures within the same plane and selecting constituent elements. The anisotropic, electrical, and mechanical properties of borophene enable precise feature manipulation and the creation of new functionalities [27]. In 2015, researchers used physical vapor deposition (PVD) to produce atomically thin borophene by depositing borons onto a silver single-crystal surface

[28]. However, a suitable boron precursor was needed for the synthesis and development of the borophene. The researchers found diborane in commercially available borazine, which allowed them to produce high-quality borophene layers. This new method introduces a new method for examining borophene properties in van der Waals heterostructures and devices [29].

4.1 Bottom-Up Versus Top-Down Approaches

The development of 2D boron nanostructures (borophene, BN nanosheets, etc.) involves two main approaches: bottom-up and top-down as depicted in Fig. 4. Bottom-up techniques involve building borophene from individual atoms or small molecules, gradually assembling the desired structure layer by layer. Synthesis techniques like chemical vapor deposition (CVD) and molecular beam epitaxy (MBE) offer advantages such as high purity, precise control over thickness, and scalability. However, they also face challenges in growth conditions and substrate interactions [6]. Top-down approaches involve breaking down bulk boron-containing materials to obtain atomically thin borophene sheets. Techniques like micromechanical cleavage, liquid exfoliation, and chemical exfoliation offer advantages such as simpler setups, higher yields, and less control over thickness, lateral size, and crystallinity. However, they also have the potential for defects or contamination [30]. Choosing the right approach depends on the desired properties and application of the borophene. For high-purity and well-defined borophene, bottom-up techniques like CVD with careful optimization might be preferred. For large quantities of borophene and acceptable structural variation, top-down exfoliation techniques could be a viable option [31, 32].

4.2 Chemical Vapor Deposition and Exfoliation Techniques

CVD is a promising technique for synthesizing 2D boron nanostructures, or borophene, with unique electronic and physical properties. These boron sheets are used in nanoelectronics, sensors, and catalysis. However, CVD faces challenges due to boron's high melting point and the need for precise control over crystal structure. Boron also interacts with common CVD substrates, affecting the resulting borophene layer's morphology. CVD growth strategies include low-temperature CVD, surface modification with a sacrificial layer, and plasma-enriched CVD. Advantages of CVD include scalability, cost-effectiveness, and precise control over thickness. Compared to molecular beam epitaxy, CVD offers a more cost-effective approach. Recently, Mazaheri et al. successfully demonstrated the creation of atomic-thickness borophene sheets using CVD with diborane (B_2H_6) pyrolysis as a pure boron source. The layers of CVD-borophene have an average thickness of 4.2 Å, possess a $\chi 3$ crystalline

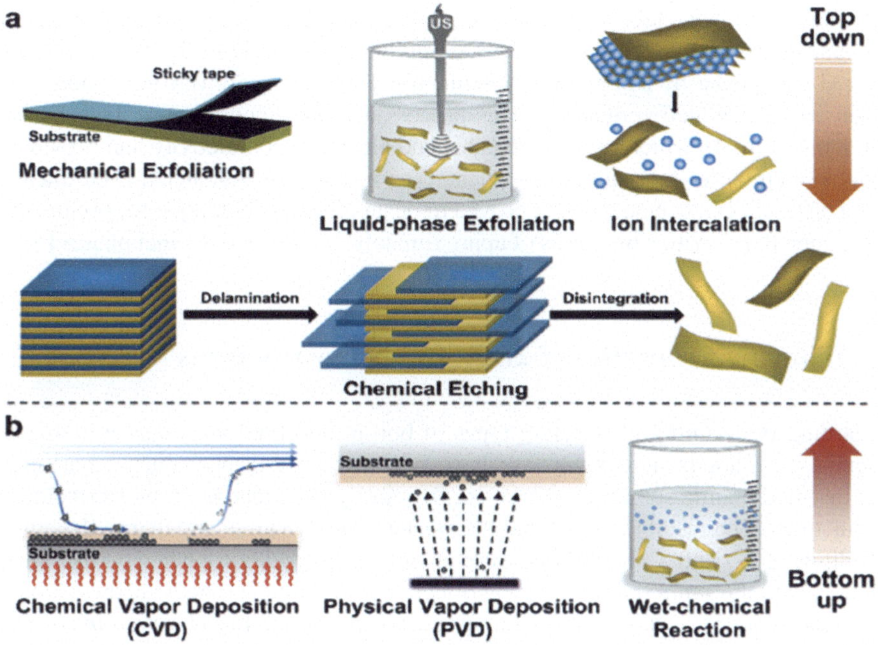

Fig. 4 Synthesis approaches of 2D boron nanostructures. Reprinted (adapted) with permission from [37]. Copyright 2021 Royal Society of Chemistry

structure, and demonstrate metallic conductivity. Additionally, they provide experimental proof that supports the creation of stacked bilayer and trilayer borophene sheets [33].

Exfoliation techniques are an alternative method for synthesizing 2D boron nanostructures, which involve separating atomically thin layers from a bulk crystal. Borophene exhibit strong interlayer bonding, making it difficult to separate individual layers without damaging the structure. Additionally, borophene can interact with the substrate surface, leading to defects or altering their properties. Exfoliation techniques often yield borophene with varying lateral sizes, which can affect their properties and device applications [34]. Common exfoliation techniques include micromechanical cleavage, liquid exfoliation, chemical exfoliation, and electrochemical exfoliation. Micromechanical cleavage involves repeated exfoliation of bulk BN crystals using adhesive tape, while liquid exfoliation involves sonicating bulk boron carbide or other boron-containing precursors in a solvent. For instance, Chahal et al. synthesized mono- and few layered borophene of device quality via micromechanical exfoliation [35]. Chemical exfoliation selectively removes specific layers from a layered boron precursor material but can introduce defects or contaminate the final product. Electrochemical exfoliation uses an electric current to promote the exfoliation of individual borophene layers from the electrode material, offering control over thickness and scalability. For instance, Sielicki and his team successfully produced few-layered borophene by using Electrochemical exfoliation to convert nonconductive boron into a conductive material. This process involved incorporating bulk boron into a metal mesh, which resulted in enhanced electrical conductivity and provided a viable method for fabricating borophene. The researchers conducted experiments using different electrolytes, which proved to be an effective technique for producing borophene flakes with a thickness of approximately 3–6 nm and distinct phases [36].

5 Tailoring Properties of 2D Boron Nanostructures

Regarding the stability of different types of boron- and borophene-based systems, scientists are not in agreement. We have investigated the mechanical, dynamic, and thermodynamic stabilities of striped, β_{12}, and χ_3 borophene by first-principles calculations. The results showed that the mechanical, dynamic, and thermodynamic stabilities of β_{12} and χ_3 borophene were favorable. Striped borophene, on the other hand, lacked the same stability. Strong directional bonding and high stiffness along the same direction account for the high stiffness and instability observed in striped borophene [11]. Different arrangements of two-dimensional boron sheets from three-dimensional boron bulk lead to distinct electrical properties. The two main categories for 3D boron solids with different kinds of B_{12} cages are semiconductors and insulators [30].

Defects have an impact on the optical, electrical, magnetic, chemical, vibrational, and thermal properties of 2D materials. In the field of electronics, particularly in transportation, grain quality and defects have the most significant impact on large-area 2D materials. However, research indicates that the presence of defects and a disordered crystal structure can result in the development of electrical, electronic, and optical technologies that are fundamentally new and highly efficient.

5.1 Doping and Functionalization Strategies

The efficacy and adaptability of various doping strategies used to adjust the band structure and change the properties of 2D materials have been well established. Interstitial doping and alloying within the crystal lattice, as well as surface doping via physisorption or chemisorption, will become important in the development of multifunctional 2D electronic hardware and optoelectronic devices that rely on electromagnetic fields. Combining different doping techniques allows one to address both the benefits and shortcomings of current systems for doping 2D materials. Many dopant atoms with potentially interesting electrical and attractive properties have not yet been thoroughly investigated for substitutional lattice doping [38]. Doping borophene can exert an influence on conductivity, much like bandgap engineering. Although n-type doping, which involves adding electron donors, can enhance conductivity, the development of p-type doping, which involves adding hole donors, is still ongoing [39].

Considering the wide variety of functionalization techniques and potential applications of the recently identified 2D boron nanosheets. The term "functionalization" here refers to the process of adding heterostructures to boron nanosheets, which modifies their chemical and physical properties. To be more exact, adding chemical groups, dopants, and strongly interacting composites is the process of functionalization [40, 41]. Recently, Ahmad and his team investigated the electrochemically enhanced nanozymatic activity demonstrated by functionalized borophene with 3,3′,5,5′-tetramethylbenzidine (TMB), and H_2O_2. In this study, the surface of the pristine borophene was initially altered by introducing thiocyanate groups to enhance the scavenging ability of hydroxyl radicals (•OH). Subsequently, the electrochemical potential applied resulted in an increased rate of oxidation reaction for TMB. Both features greatly enhanced the detection limit and substantially reduced the detection time [42].

5.2 *Engineering Bandgaps and Electrical Conductivity*

Several strategies for bandgap engineering include doping, which involves the introduction of foreign atoms (dopants) into the borophene lattice, which results in the creation of energy levels inside the band structure. This, in turn, leads to the formation of a bandgap. This can be accomplished by substitutional doping involving the replacement of boron atoms with elements such as nitrogen or carbon. Interstitial doping involves the insertion of dopant atoms into the empty spaces within the borophene lattice. Recently, Wang et al. produced hydroxyl-functionalized borophene (borophene-OH) with an adjustable bandgap using the process of liquid-phase exfoliation in a 2-butanol solvent. Boron and 2-butanol were used to create smooth borophene by matching their surface energies. The exfoliation process resulted in the formation of unsaturated B sites through the breaking of B–B bonds. These sites coordinated with OH groups to produce semiconducting borophene-OH. By adjusting the thickness of the borophene, the bandgap could be varied between 0.65 and 2.10 eV [43]. Strain engineering is the application of mechanical strain to the borophene sheet, which can modify its electrical band structure and perhaps create a band gap. This can be accomplished with methodologies such as:

- **Substrate Engineering:** Employing substrates with a lattice mismatch to generate strain on the borophene layer. External strain devices involve the application of precise mechanical pressure on the borophene utilizing micro/nano-mechanical devices.
- **Heterostructure Formation:** The combination of borophene with other two-dimensional materials can generate novel electrical interactions at the interface, resulting in the creation of a bandgap.

This can encompass vertical heterostructures, which involve the arrangement of borophene with other 2D materials such as h-BN or transition metal dichalcogenides (TMDCs). Lateral heterostructures include the formation of in-plane connectors between distinct borophene polymorphs or other 2D materials [44].

Defect Engineering involves deliberately introducing controlled flaws into the borophene lattice to change its conductivity. However, precise regulation is essential to prevent the formation of undesired scattering centers that diminish conductivity. Chemical functionalization involves the process of attaching functional groups to the surface of borophene, which can modify its electrical properties and conductivity. This can be utilized to customize the material for particular uses [45].

6 Applications of Bulk Boron and 2D Boron Nanostructures

6.1 Potential Applications in Electronics, Sensors, and Energy Storage

Boron is essential in energy-related research due to its involvement in a variety of applications, including the storage of chemical and electrical energy, the activation and synthesis of energy-rich small molecules, and the conversion of electrical energy into light. The unique characteristics of boron, such as its electron deficiency and the presence of an unoccupied p orbital, are the primary reasons for these applications. All of these attributes facilitate the formation of a diverse array of compounds that possess a variety of physical and chemical properties. Consequently, boron-containing compounds possess distinctive physicochemical properties that suggest they have substantial potential for use in a variety of disciplines, including gas storage, optical data storage, waveguides, multiphoton microscopy, energy conversion and storage, molecular machines, and even cell biology [46]. 1D boron nanostructures possess two primary physical characteristics: electronic transport and field emission. These properties make them highly intriguing due to their significant potential in various applications such as field emission, flexible displays, superconductors, energy storage, and nanoelectronic devices. Nevertheless, despite their popularity, numerous crucial scientific challenges remain that need to be comprehended to fully utilize these 1D nanomaterials. An important factor to consider is the correlation between the transport characteristics and crystal structures of boron. Boron exhibits multiple distinct crystal structures. Studying the transport properties of boron nanostructures with various crystal structures is crucial. Furthermore, it is imperative to allocate additional resources to comprehend the transportation mechanism by manipulating measurement conditions, including temperature, strain, and doping. Field emission applications require careful consideration of both synthesis procedures and property measurements. Uniformly sized boron nanostructures with appropriate density and well-controlled spacing are crucial for enhancing the field emission properties. Furthermore, the demand for flexible and conductive substrates for cultivating boron nanostructures is crucial in achieving flexible displays [47]. 2D borophene or borophane exhibits unique advantages in flexible energy conversion devices or high-speed electronic devices due to its metallic properties, bigger specific surface area, higher Young's modulus, and exceptionally high Fermi velocity [48]. Additionally, 1D nanostructures composed of boron have pronounced light-absorption characteristics and prompt photocurrent reactions, indicating their suitability as viable options for flexible photodetectors operating within the visible and infrared spectrums [32].

The suitability of a material as an electrode material depends on its larger capacitance, excellent electrical conductivity, and ionic conductivity. Borophene, being a 2D substance, exhibits a strong surface reactivity, which facilitates the achievement of exceptionally high storage capacity in metallic materials. Furthermore, the metallic band structure of borophene enhances electrical conductivity, making it suitable for use as an electrode in metal-ion batteries. Jena et al. reported the synthesis of borophene at Ag substrate under ultra-high vacuum conditions [8]. Borophene-based materials have been effectively utilized as electrode materials in high-performance supercapacitors and rechargeable innovative batteries [48]. Furthermore, borophene-based materials, which possess a significant quantity of reactive sites, have proven to be very effective catalysts in facilitating the hydrogen evolution reaction (HER) and oxygen evolution reaction (OER). For instance, Saad et al. discovered that borophene, when adorned with Ag nanoparticles (Ag@B), acts as a support and significantly improves the performance of Co_3O_4 catalysts in the OER. The presence of conductive Ag-nanoparticles on borophene promotes the formation of thin Co_3O_4 nanoplates through an epitaxy model, leading to a significant increase in the number of Co_3O_4 catalyst sites on the surface. Meanwhile, the presence of Ag nanoparticles enhances the electrical conductivity of the support, enabling the transfer of electrons from Co^{III} atoms to create catalytically active Co^{IV} sites. Furthermore, the B atoms exhibit a significant interaction with the Co and O atoms present on the Co_3O_4 nanoplates. This interaction effectively lowers the energy barrier associated with the rate-determining step of the OER and enhances the stability of the borophene supports [49]. These materials have also been utilized in the construction of an innovative hydroelectric generator designed to harness energy from water. In addition, borophene and related heterostructures have demonstrated exceptional detecting capabilities in humidity sensors, gas sensors, pressure sensors, and photodetectors. In addition, researchers have successfully produced nonvolatile memory devices using borophene, which exhibits a low set voltage, a high ON/OFF ratio, excellent stability, and even a wide light response [50].

6.2 Biomedical Applications in Biosensing, Bioimaging, and Drug Delivery

Borophene is a highly prospective substance for theranostic and drug delivery applications due to its fluorescence and photoacoustic contrast, as well as its favorable material characteristics. Furthermore, borophene exhibits therapeutic properties for photothermal therapy (PTT) and photodynamic therapy (PDT). Recently, Qi et al. investigated the potential role of borophene as a radiosensitizer in boron neutron capture therapy (BNCT) and particle therapy (PT) [51]. It is possible to modify borophene by employing fluorescent molecules. In laboratory settings or within living organisms, fluorophores can be affixed or incorporated onto the surface of borophene to facilitate fluorescence imaging. Additionally, Ma et al. developed ultrathin boron nanosheets using liquid exfoliation, resulting in nanosheets that exhibit excellent photostability and favorable biocompatibility. These boron nanosheets can be used as a new and efficient fluorescent probe for imaging cells [52]. It is possible to monitor the biological activities of borophene-based nanosystems, including their absorption pathways and cellular localization, through imaging research. The precise location of tumors, the precise cultivation of hazardous tumor cells, and the mitigation of adverse consequences are all possible with this data. In the biomedical domain, the utilization of borophene-based diagnosis-guided therapy is promising [53]. Additionally, it is imperative to broaden the application of borophene nanomaterials in other scientific disciplines, including 0D, 1D, 2D, and 3D. Furthermore, the biosensing capabilities of borophene can be broadened to include a diverse array of biomolecules, enabling the detection and analysis of a variety of analytes. This has substantial implications for the diagnosis and prognosis of numerous diseases. However, the initial phases of research are still underway regarding the production of borophene and its potential for use in biosensing. These materials have been explored for their use in biosensors, where their high sensitivity and selectivity make them ideal for detecting biomarkers and monitoring biological processes. The applications of borophene nanomaterials in biosensors include glucose detection [54], DNA sequencing [55], and cancer diagnosis, among others. The unique characteristics of borophene necessitate the ongoing discovery and investigation of new synthetic techniques. This would expand the scope of their implementations in established scientific domains such as NEMS and MEMS [2]. Borophene has recently garnered attention in the field of cancer diagnosis and treatment due to its exceptional properties, which include an exceptionally large specific surface area, outstanding biocompatibility, a high degree of efficiency in converting light into heat, and beneficial effects on bioimaging [56]. Borax and its derivatives have demonstrated a substantial influence on human health throughout history. The potential for boron-related products to have a significant impact on the field of biomedicine is highly probable. As a result, borophene likely possesses exceptional immune functionality, including cytokine-like activity, enzyme-like activity, and agonist activity that is comparable to that of oxidized graphene [57]. Borophene is a promising candidate for use as a drug delivery platform in the field of immunotherapy due to these attributes. In

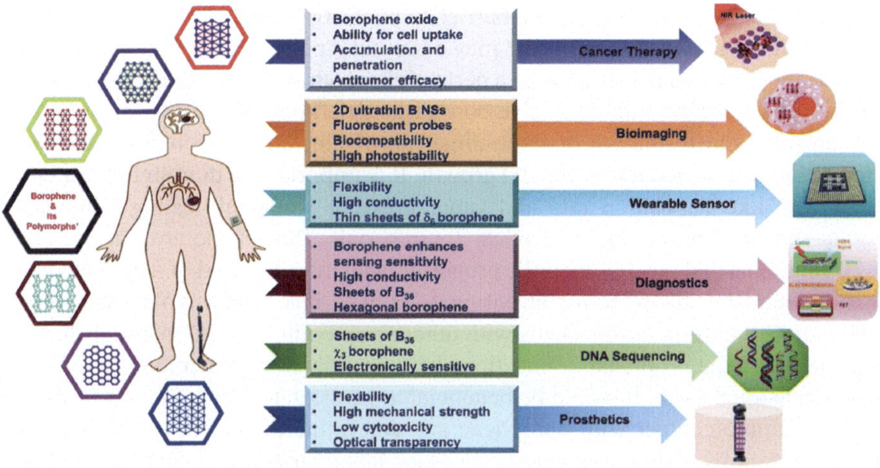

Fig. 5 Graphical representation of potential biomedical applications of borophene. Reprinted (adapted) with permission from [59]. Copyright 2022 Royal Society of Chemistry

comparison to other boron-containing compounds, borophene has superior boron purity, rendering it a promising candidate for the delivery of boron in the treatment of glioma. Additionally, it possesses exceptional tumor targeting and photothermal properties. Borophene may also demonstrate specific enzymatic activity due to its 2D structure and abundance of surface active sites [58]. The biological applications of 2d borophene are summarized in Fig. 5.

6.3 *Future Directions and Challenges*

Ensuring the precise and deliberate production of abundant, high-quality samples is essential for achieving any potential industrial uses of 2D boron nanoclusters. For this purpose, the CVD technique is still the preferred option. Nevertheless, numerous endeavors employing this approach yielded formless configurations, clusters, or dense coatings. Through meticulous analysis of their growth conditions and precursors, researchers offer multiple formulas that could facilitate the CVD production of borophene. The initial crucial factor is the temperature at which growth occurs. Elevated temperature can initiate nucleation through the three-dimensional pathway, leading to the formation of boron films. In contrast, lower temperatures cannot surpass the necessary barrier for two-dimensional development, resulting in the creation of non-crystalline clusters. Therefore, it is crucial to precisely regulate the growth temperature, ensuring that it is sufficient to initiate 2D nucleation while preventing 3D nucleation. Furthermore, it is necessary to utilize pure boron as the precursor. Additional components present in the precursors, such as hydrogen and oxygen, can stabilize the three-dimensional nucleus by neutralizing exposed bonds on its substrates, thus reducing the barrier to nucleation. According to a study, the presence of oxygen in the precursors may contribute to the yield of boron nanofilms. Furthermore, a metal substrate with a perfectly smooth surface at the atomic level is highly suitable for creating extensive boron islands. The presence of rough surfaces may enhance the formation of three-dimensional clusters by increasing the local interaction area, hence inhibiting 2D growth. It is important to thoroughly test other parameters like deposition rate and vapor pressure when fabricating conjugated atomic sheets of borophene and other 2D materials. Borophene are a novel class of 2D materials that possess metallic properties and are composed of a light element. This distinguishes them from conventional 2D materials, which typically exhibit a band gap. Combining boron sheets with other semiconducting 2D materials has the potential to create co-planar connections between metals and semiconductors in a two-dimensional form. It would be beneficial to investigate which 2D materials can establish ohmic contact with borophene and determine the extent of the Schottky barrier formed by different materials. In-plane heterostructures featuring an ohmic contact enable the seamless integration of semiconducting channels and electrodes in a single material. On the other hand, heterostructures with a Schottky barrier can be utilized to create atomically thin diodes and transistors [11]. Identifying the phase structure of elemental boron has become a highly active field of study, however, the precise position of the phase boundaries remains unknown. The primary difficulties in this field arise from the inherent characteristics of elemental boron, such as its intricate crystal structures in different forms that are hindered by significant barriers to change and its structural and electrical properties that strongly favor the inclusion of impurities. The significant kinetic barrier poses a substantial challenge to the experimental determination of phase boundaries. Due to sluggish kinetics, the direct correlation between phase boundaries in high-pressure and temperature studies is unsuccessful at low temperatures and pressures [60].

7 Conclusion

This chapter has explored the fundamental building blocks of boron, both in its three-dimensional bulk crystal form and in exciting new 2D nanostructures. We have seen that bulk boron crystals exhibit a variety of fascinating structures, each with unique properties. The understanding of these structures is crucial for comprehending the physical and chemical behavior of bulk boron. The advent of 2D boron nanostructures, such as borophene, has opened a new chapter in boron research. These atomically thin sheets possess remarkable electronic and mechanical properties that hold immense promise for future technological advancements. By delving into the structural units of both bulk and 2D boron, this chapter has laid the foundation for further exploration of their properties and potential applications. Future research directions could involve investigating the manipulation of these structures to tailor their functionalities to specific technological needs. Additionally, exploring the interplay between the structures and properties of bulk and 2D boron holds vast potential for the development of novel materials with unprecedented characteristics. In conclusion, understanding the structural units of boron, from bulk crystals to 2D wonders, is essential for unlocking the full potential of this versatile element and paving the way for groundbreaking discoveries in materials science and beyond.

Acknowledgements The Director of CSIR-AMPRI, Bhopal, is gratefully acknowledged by all authors for his invaluable support of this endeavor. S.Y. is grateful to the CSIR for providing financial assistance in the form of a senior research fellowship (SRF).

References

1. Glavin, N.R., et al.: Emerging applications of elemental 2D materials. Adv. Mater. **32**(7), 1–22 (2020). https://doi.org/10.1002/adma.201904302
2. Gupta, G.H., Kadakia, S., Agiwal, D., Keshari, T., Kumar, S.: Borophene nanomaterials: synthesis and applications in biosensors. Mater. Adv. **5**(5), 1803–1816 (2024). https://doi.org/10.1039/D3MA00829K
3. Zhu, Y., Cai, J., Hosmane, N.S., Zhang, Y.: Chapter 1—Introduction: basic concept of boron and its physical and chemical properties. In: Zhu Y. (ed.) Developments in Inorganic Chemistry. Fundamentals and Applications of Boron Chemistry, vol. 2, pp. 1–57. Elsevier (2022). ISBN 9780128221273. https://doi.org/10.1016/B978-0-12-822127-3.00003-X
4. Kong, L., Wu, K., Chen, L.: Recent progress on borophene: growth and structures. Front. Phys. (Beijing) **13**(3), 138105 (2018). https://doi.org/10.1007/s11467-018-0752-8
5. Boustani, I., Quandt, A.: Nanotubules of bare boron clusters: Ab initio and density functional study. Europhys. Lett. **39**(5), 527–532 (1997). https://doi.org/10.1209/epl/i1997-00388-9
6. Ayodhya, D., Veerabhadram, G.: A brief review on synthesis, properties and lithium-ion battery applications of borophene. FlatChem **19**, 100150 (2020). https://doi.org/10.1016/j.flatc.2019.100150
7. Kostoglou, N., Polychronopoulou, K., Rebholz, C.: Thermal and chemical stability of hexagonal boron nitride (h-BN) nanoplatelets. Vacuum **112**, 42–45 (2015). https://doi.org/10.1016/j.vacuum.2014.11.009
8. Xie Z. et al.: Two-dimensional borophene: properties, fabrication, and promising applications. Research 2020 (2024). https://doi.org/10.34133/2020/2624617
9. Yang, M., Jin, H., Sun, Z., Gui, R.: Monoelemental two-dimensional boron nanomaterials beyond theoretical simulations: from experimental preparation, functionalized modification to practical applications. Adv. Colloid Interface Sci. **304**, 102669 (2022). https://doi.org/10.1016/j.cis.2022.102669
10. Xie, S.Y., Wang, Y., Bin Li, X.: Flat boron: a new cousin of graphene. Adv. Mater. **31**(36), 1–13 (2019). https://doi.org/10.1002/adma.201900392
11. Zhang, Z., Penev, E.S., Yakobson, B.I.: Two-dimensional boron: structures, properties and applications. Chem. Soc. Rev. **46**(22), 6746–6763 (2017). https://doi.org/10.1039/C7CS00261K
12. Hayami, W., Hiroto, T., Soga, K., Ogitsu, T., Kimura, K.: Thermodynamic stability of elemental boron allotropes with varying numbers of interstitial atoms. J. Solid State Chem. **329**, 124407 (2024). https://doi.org/10.1016/j.jssc.2023.124407
13. Mannix, A.J., Zhang, Z., Guisinger, N.P., Yakobson, B.I., Hersam, M.C.: Borophene as a prototype for synthetic 2D materials development. Nat. Nanotechnol. **13**(6), 444–450 (2018). https://doi.org/10.1038/s41565-018-0157-4
14. Sun, X., et al.: Two-dimensional boron crystals: structural stability, tunable properties, fabrications and applications. Adv. Funct. Mater. **27**(19), 1603300 (2017). https://doi.org/10.1002/adfm.201603300

15. Widom, M., Mihalkovič, M.: Symmetry-broken crystal structure of elemental boron at low temperature. Phys. Rev. B **77**(6), 64113 (2008). https://doi.org/10.1103/PhysRevB.77.064113
16. Shirai, K., Dekura, H., Mori, Y., Fujii, Y., Hyodo, H., Kimura, K.: Structural study of α-rhombohedral boron at high pressures. J. Phys. Soc. Jpn. **80**(8), 84601 (2011). https://doi.org/10.1143/JPSJ.80.084601
17. Li, D., et al.: High-pressure close-packed structure of boron. RSC Adv. **4**(1), 203–207 (2014). https://doi.org/10.1039/C3RA45777J
18. He, C., Zhong, J.X.: Structures, stability, mechanical and electronic properties of α-boron and α*-boron. AIP Adv. **3**(4), 42138 (2013). https://doi.org/10.1063/1.4804138
19. Cherednichenko, K.A., Solozhenko, V.L.: Thermal expansion of α-boron and some boron-rich pnictides. Solid State Commun. **303–304**, 113735 (2019). https://doi.org/10.1016/j.ssc.2019.113735
20. Masago, A., Shirai, K., Katayama-Yoshida, H.: Crystal stability of α-and β-boron. Phys. Rev. B—Conden. Matter Mater. Phys. **73**(10), 104102 (2006)
21. Albert, B., Hillebrecht, H.: Boron: elementary challenge for experimenters and theoreticians. Angew. Chem. Int. Ed. **48**(46), 8640–8668 (2009). https://doi.org/10.1002/anie.200903246
22. Li, D., et al.: From two- to three-dimensional van der Waals Layered structures of boron crystals: an ab initio study. ACS Omega **4**(5), 8015–8021 (2019). https://doi.org/10.1021/acsomega.9b00534
23. Jian, T., Chen, X., Li, S.-D., Boldyrev, A.I., Li, J., Wang, L.-S.: Probing the structures and bonding of size-selected boron and doped-boron clusters. Chem. Soc. Rev. **48**(13), 3550–3591 (2019). https://doi.org/10.1039/C9CS00233B
24. Sergeeva, A.P., et al.: Understanding boron through size-selected clusters: structure, chemical bonding, and fluxionality. Acc. Chem. Res. **47**(4), 1349–1358 (2014)
25. Piazza, Z.A., Hu, H.-S., Li, W.-L., Zhao, Y.-F., Li, J., Wang, L.-S.: Planar hexagonal B36 as a potential basis for extended single-atom layer boron sheets. Nat. Commun. **5**(1), 3113 (2014). https://doi.org/10.1038/ncomms4113
26. Tai, G., et al.: Synthesis of atomically thin boron films on copper foils. Angew. Chem. Int. Ed. **54**(51), 15473–15477 (2015). https://doi.org/10.1002/anie.201509285
27. Khan, K., et al.: Synthesis, properties and novel electrocatalytic applications of the 2D-borophene Xenes. Prog. Solid State Chem. **59**, 100283 (2020). https://doi.org/10.1016/j.progsolidstchem.2020.100283
28. Tsai, H.S., Liang, J.H.: Production and potential applications of elemental two-dimensional materials beyond graphene. ChemNanoMat **3**(9), 604–613 (2017). https://doi.org/10.1002/cnma.201700096
29. Cuxart, M.G., et al.: Borophenes made easy. Sci. Adv. **7**(45), eabk1490 (2024). https://doi.org/10.1126/sciadv.abk1490
30. Li, D., et al.: 2D boron sheets: structure, growth, and electronic and thermal transport properties. Adv. Funct. Mater. **30**(8), 1904349 (2020). https://doi.org/10.1002/adfm.201904349
31. Mishra, R.K., Sarkar, J., Verma, K., Chianella, I., Goel, S., Nezhad, H.Y.: Borophene: A 2D wonder shaping the future of nanotechnology and materials science. Nano Mater. Sci. (2024). https://doi.org/10.1016/j.nanoms.2024.03.007
32. Tian, Y., et al.: Inorganic boron-based nanostructures: synthesis, optoelectronic properties, and prospective applications. Nanomaterials **9**(4) (2019). https://doi.org/10.3390/nano9040538.
33. Mazaheri, A., Javadi, M., Abdi, Y.: Chemical vapor deposition of two-dimensional boron sheets by thermal decomposition of diborane. ACS Appl. Mater. Interfaces **13**(7), 8844–8850 (2021). https://doi.org/10.1021/acsami.0c22580
34. Zhang, F., et al.: Few-layer and large flake size borophene: preparation with solvothermal-assisted liquid phase exfoliation. RSC Adv. **10**(46), 27532–27537 (2020). https://doi.org/10.1039/D0RA03492D
35. Chahal, S., et al.: Borophene via micromechanical exfoliation. Adv. Mater. **33**(34), 1–13 (2021). https://doi.org/10.1002/adma.202102039
36. Sielicki, K., Maślana, K., Chen, X., Mijowska, E.: Bottom up approach of metal assisted electrochemical exfoliation of boron towards borophene. Sci. Rep. **12**(1), 15683 (2022). https://doi.org/10.1038/s41598-022-20130-w

37. Huang, H., Feng, W., Chen, Y.: Two-dimensional biomaterials: material science, biological effect and biomedical engineering applications. Chem. Soc. Rev. (2021). https://doi.org/10.1039/d0cs01138j
38. Alam, S., Asaduzzaman Chowdhury, M., Shahid, A., Alam, R., Rahim, A.: Synthesis of emerging two-dimensional (2D) materials—advances, challenges and prospects. FlatChem **30**, 100305 (2021). https://doi.org/10.1016/j.flatc.2021.100305
39. Liu, X., Wang, L., Yakobson, B.I., Hersam, M.C.: Nanoscale probing of image-potential states and electron transfer doping in borophene polymorphs. Nano Lett. **21**(2), 1169–1174 (2021). https://doi.org/10.1021/acs.nanolett.0c04869
40. Tang, X., Gu, J., Shang, J., Chen, Z., Kou, L.: Double-sided surface functionalization: an effective approach to stabilize and modulate the electronic structure of graphene-like borophene. InfoMat **3**(3), 327–336 (2021). https://doi.org/10.1002/inf2.12126
41. Han, C., et al.: 2D boron nanosheet architectonics: opening new territories by smart functionalization. J. Mater. Chem. A Mater. **10**(6), 2736–2750 (2022). https://doi.org/10.1039/D1TA09562E
42. Ahmed, S.R., et al.: Electrochemical assisted enhanced nanozymatic activity of functionalized borophene for H2O2 and tetracycline detection. Biosens. Bioelectron. **246**, 115857 (2024). https://doi.org/10.1016/j.bios.2023.115857
43. Wang, X., et al.: Bandgap engineering of hydroxy-functionalized borophene for superior photoelectrochemical performance. Angew. Chem. Int. Ed. **59**(52), 23559–23563 (2020). https://doi.org/10.1002/anie.202010723
44. Wei, J., Li, W., Liao, B., Bian, B.: Electronic and optical properties of vertical borophene/MoS2 heterojunctions. Mater. Chem. Phys. **252**(2019), 123305 (2020). https://doi.org/10.1016/j.matchemphys.2020.123305.
45. Pham, V.T., Fang, T.H.: Anisotropic mechanical strength, negative Poisson's ratio and fracture mechanism of borophene with defects. Thin Solid Films **709**, 138197 (2020). https://doi.org/10.1016/j.tsf.2020.138197
46. Ali, F.: Chapter 4—Boron materials for energy applications. In: Zhu, Y. (ed.) Developments in Inorganic Chemistry. Fundamentals and Applications of Boron Chemistry, vol. 2, pp. 203–289. Elsevier (2022). ISBN 9780128221273. https://doi.org/10.1016/B978-0-12-822127-3.00004-1
47. Tian, J., Xu, Z., Shen, C., Liu, F., Xu, N., Gao, H.-J.: One-dimensional boron nanostructures: prediction, synthesis, characterizations, and applications. Nanoscale **2**(8), 1375–1389 (2010). https://doi.org/10.1039/C0NR00051E
48. Jena, N.K., Araujo, R.B., Shukla, V., Ahuja, R.: Borophane as a benchmate of graphene: a potential 2D material for anode of Li and Na-ion batteries. ACS Appl. Mater. Interfaces **9**(19), 16148–16158 (2017)
49. Saad, A., et al.: Ag nanoparticles modified crumpled borophene supported Co3O4 catalyst showing superior oxygen evolution reaction (OER) performance. Appl. Catal. B **298**, 120529 (2021). https://doi.org/10.1016/j.apcatb.2021.120529
50. Hou, C., Tai, G., Liu, Y., Wu, Z., Liang, X., Liu, X.: Borophene-based materials for energy, sensors and information storage applications. Nano Res. Energy **2**, e9120051 (2023). https://doi.org/10.26599/NRE.2023.9120051
51. Qi, P., et al.: The potential role of borophene as a radiosensitizer in boron neutron capture therapy (BNCT) and particle therapy (PT). Biomater Sci. **8**(10), 2778–2785 (2020). https://doi.org/10.1039/d0bm00318b
52. Ma, D., et al.: Ultrathin boron nanosheets as an emerging two-dimensional photoluminescence material for bioimaging. Nanoscale Horiz **5**(4), 705–713 (2020). https://doi.org/10.1039/c9nh00698b
53. Duo, Y., et al.: Borophene-based biomedical applications: status and future challenges. Coord. Chem. Rev. **427**, 213549 (2021). https://doi.org/10.1016/j.ccr.2020.213549
54. Taşaltın, C., Türkmen, T.A., Taşaltın, N., Karakuş, S.: Highly sensitive non-enzymatic electrochemical glucose biosensor based on PANI: β12 Borophene. J. Mater. Sci.: Mater. Electron. **32**(8), 10750–10760 (2021). https://doi.org/10.1007/s10854-021-05732-w

55. Rastgou, A., Soleymanabadi, H., Bodaghi, A.: DNA sequencing by borophene nanosheet via an electronic response: a theoretical study. Microelectron. Eng. **169**, 9–15 (2017). https://doi.org/10.1016/j.mee.2016.11.012
56. Ji, X., et al.: A Novel top-down synthesis of ultrathin 2D boron nanosheets for multimodal imaging-guided cancer therapy. Adv. Mater. **30**(36), 1803031 (2018). https://doi.org/10.1002/adma.201803031
57. Han, M., et al.: Protein corona and immune responses of borophene: a comparison of nanosheet-plasma interface with graphene and phosphorene. ACS Appl. Bio Mater. **3**(7), 4220–4229 (2020). https://doi.org/10.1021/acsabm.0c00306
58. Ou, M., et al.: The Emergence and Evolution of Borophene. Adv. Sci. **8**(12), 2001801 (2021). https://doi.org/10.1002/advs.202001801
59. Yadav, S., Sadique, M.A., Kaushik, A., Ranjan, P., Khan, R., Srivastava, A.K.: Borophene as an emerging 2D flatland for biomedical applications: current challenges and future prospects. J. Mater. Chem. B **10**, 8 (2022). https://doi.org/10.1039/d1tb02277f
60. Ogitsu, T., Schwegler, E., Galli, G.: β-rhombohedral boron: at the crossroads of the chemistry of boron and the physics of frustration. Chem. Rev. **113**(5), 3425–3449 (2013). https://doi.org/10.1021/cr300356t

Advantages of 2D Boron Nanosheets Over Other 2D Nanomaterials

Deepak Kumar, Pashupati Pratap Neelratan, Anshika Gupta, and Sanjeev K. Sharma

Abstract Two-dimensional (2D) nanomaterials, including graphene (Gr), have attracted significant interest for a variety of applications, such as biomedicine, energy harvesting/storage, sensing, and catalysis. The high mechanical stiffness, excellent electrical conductivity, optical transparency, and biocompatibility have led to their rapid advances. In addition to Gr, other 2D nanomaterials, such as boron (B) and their derivatives, including 2D-boron nanosheets (2D-BNSs), have shown better physical, electrical, and mechanical properties with high in-plane stiffness, strength, and flexibility. This chapter focuses on the physical, electrical, optical, and mechanical properties, the structural advantages of 2D boron nanosheets, and their derivatives compared to other 2D materials.

Keywords 2D Boron nanosheets · Borophene · Electrical and optical properties

1 Introduction

Nanostructured 2D materials have attracted wide attention for academics and industrial purposes since graphene (Gr) was discovered in 2004 [1]. Nowadays, the term "2D nanomaterials" refers to over a hundred distinct materials in the family of carbon (C) nanomaterials: Gr, carbon nanotubes (CNTs), fullerenes (C_{60}), graphene QDs, graphene oxide (GO), graphitic carbon nitride (g-C_3N_4) and other variants [2]. Compared to other materials, 2D boron nanosheets (2D-BNSs) are introduced as a new member of 2D materials and have been investigated extensively [3]. 2D-hexagonal boron nitride (2D-hBN) is also known as the white Gr, showing better physical, electrical, and mechanical properties. The transition metal dichalcogenides (TMDs), such as MoS_2, $MoSe_2$, WS_2, and WSe_2, transition metal oxides (TMOs), metal–organic frameworks (MOFs), and homoatomic materials (antimonene, silicene, germanene, phosphorene) have also been investigated [4]. Figure 1

D. Kumar · P. P. Neelratan · A. Gupta · S. K. Sharma (✉)
Biomaterials and Sensor Laboratory, Department of Physics, CCS University Meerut Campus, Meerut, Uttar Pradesh 250004, India
e-mail: sksharma18@ccsuniversity.ac.in

R. Khan et al. (eds.), *2D Boron Nanosheets*, Engineering Materials,
https://doi.org/10.1007/978-981-97-7014-4_2

shows the schematic illustration of 2D nanomaterials designed for the next generation of smart devices.

Boron is an extraordinary element and is seen in the periodic table between Be (metal) and C (non-metal). Boron (B) contains three valence electrons in its outer-shell electron configuration, with a ground-state structure of $2s^2\ 2p^1$; promoting one electron from 2 s to 2p orbitals results in four accessible valence orbitals. 2D boron layers are considered unstable due to the lack of enough electrons to fill all-electron orbitals, similar to the carbon (C) systems [5]. Therefore, 2D boron nanosheets have been developed and expanded the vast family of 2D materials.

2D layered structures offer novel physicochemical characteristics, such as optical nonlinear coefficients, high in-plane stiffness, highly confined visual modes, linear polarization with little loss, strength, and incredibly low flexural rigidity, compared to bulk materials. The potent catalytic activity and quick response to external pH and temperature variation stimuli are critical features of 2D materials. The electric properties of these materials also demonstrate tuneable bandgap and improved

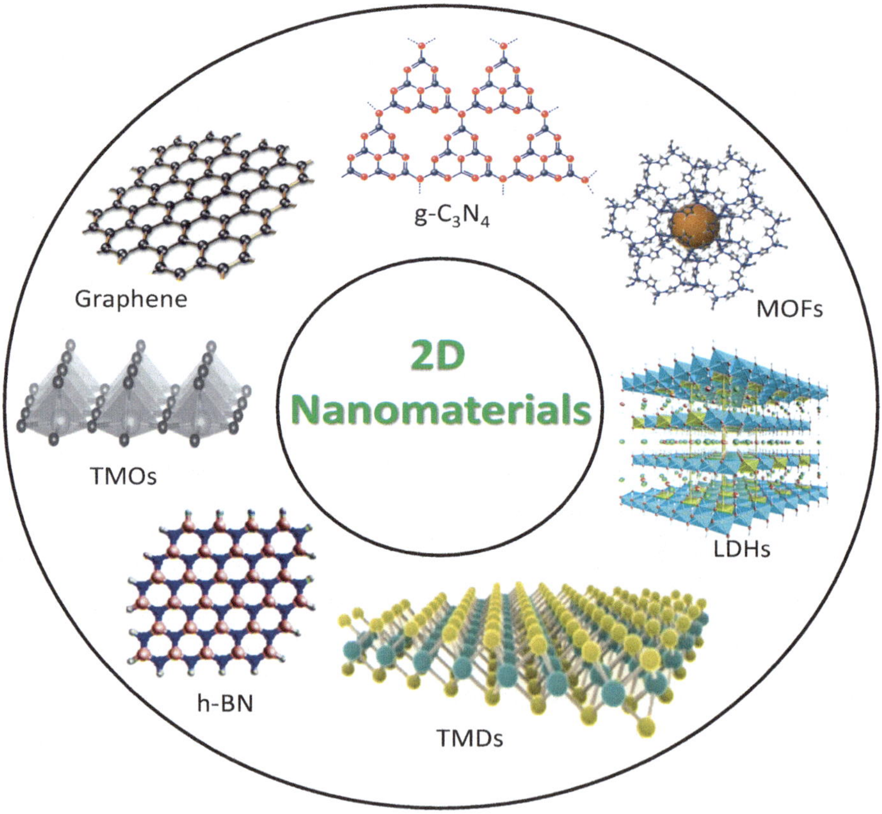

Fig. 1 The schematic illustration of 2D materials including Gr, g-C_3N_4, MOFs, LDHs, TMDs, TMOs, and BNSs

quantum transport [6]. Because of the rapid evolution of 2D materials, extensive research efforts have been directed toward developing nanocomposites composed of a supportive or template material for applications in solid-state/flexible electronics, photonics, energy storage, and biomedical (drug delivery, imaging, and biochemical/biological sensing) [7]. Synthetic and biological macromolecules functionalized with organic groups are among other complementary materials that can enhance the surface chemistry of 2D sheets.

Although the structural anisotropy makes Gr highly flexible, 2D-BNSs have record-high flexibility and a higher stiffness-to-weight ratio. It is also a good conductor of electricity and heat. The h-BN nanosheets are the most commonly used 2D-BNSs and are considered stable under standard conditions. The h-BN can be found in two-dimensional nanosheets (2D BNSs), one-dimensional nanotubes (1D BNNT), and zero-dimensional fullerenes (0D fullerenes) [8, 9]. Boron nanosheets are widely studied owing to their outstanding properties, including high surface area, visible-light transmittance, thermally conductive behavior, inert nature, extraordinary strength, and durability [10]. Unlike GO, BNSs display an electrically insulating property with fluctuating humidity and temperature. In various electrochemical energy conversion and storage devices, they are a more dependable alternative for tuning and use as electrolyte membranes [11]. Among the 2D boron nanosheets, Borophene is an emerging 2D flatland for biomedical applications. Another benefit of borophene or 2D BNSs is their adaptability due to top-down fabrication, which enables the elasticity to be reflexively tensed to form various structures. Because of their chemical functionalization, BNSs have the potential to change their physical characteristics. The effects of chemical functionalization of few-layered BNSs on charge carrier injection, chemical anchoring, and energy band structure have been reported in theoretical studies [12]. This chapter discusses the details of physical, electrical, optical and mechanical properties including the structural advantages of 2D boron nanosheets over other 2D materials.

2 Structural Properties of 2D Boron Nanosheets

2.1 Planar Boron Clusters

Boron clusters have received much attention in materials science, physical chemistry, nanoscience, and nanotechnologies over the last decade [13]. These polyhedral clusters/groups are either bare or combined with other elements. They have peculiar structural and electronic characteristics and chemical bonding due to the component's intrinsic electron deficiency. The practical and theoretical study concluded that 2D boron clusters have better properties than bulk boron [14]. Planar or nearly planar structure has been attained by B clusters, with B_{16}^+ as cations, B_{20} as neutral clusters, and B_{28} as anions. Even though the B_{28} clusters have formed, vacancies exist in the hexagonal and twin-hexagonal planar structures of the B_{30}, B_{35}, and B_{36}

clusters. These vacancies offer indirect experimental support for the feasibility of 2D atomic-thin boron sheets or borophenes [15].

Planar boron clusters are exciting materials concerning their ground states and atomic configurations for investigating 2D boron structures and their applications [16]. Figure 2 shows the structures of various boron clusters that tend to form planar or quasi-planar (2D) structures since B has an electron deficiency, where bonding electrons are shared or delocalized within the 2D plane. B compounds are generally characterized by triangles of B_3. As the size of clusters increases, such as tetragonal, pentagonal, and hexagonal, the vacancies can be seen in 2D boron clusters. At the first instance of hexagonal, the vacancy was found in B_{36}, which provided experimental support for the possibility of monolayer B with hexagonal vacancies that inspired the concept of borophenes [17]. According to photoelectron spectroscopy, one B atom is placed in the middle of the hexagonal structure of B_7 clusters [18]. When the number of atoms in the cluster is increased up to 10–15, the planar configurations of B clusters were still observed and well preserved. It is observed that all clusters of up to 15 boron atoms are made up of a triangular lattice with no hexagonal lattice. Kiran et al. investigated that the critical size of boron clusters could be found and maintain a planar configuration [19]. When the number of atoms exceeds 18, the 1D double-ring tubular structure is formed with a diameter of 5.2 Å, for example, the tubular structure of B_{20} anion is formed as an isoenergetic to 2D structure.

The charge doping of hollow hexagonal vacancies (HHs) effectively stabilizes the structures of planar boron clusters. It has been observed that the structure of B_{36} could achieve a highly stable quasi-planar (2D) structure. In the neutral state, the fullerene-like cage structure of B_{40} is shown to be the most stable form. The ground state transforms to a quasi-planar isomer upon introducing an external electron and two adjacent hexagonal holes in the ground state. Higher densities and more evenly spread HHs can lower the total energies and place planar or quasi-planar structures in their ground states for a more giant B cluster. When 2, 3, 5, 6, and 7 HHs are introduced, the B_{56}, B_{70}, B_{84}, B_{98}, B_{112}, and B_{120} clusters show perfect planarity [21]. These HHs in the boron clusters are ascribed to stabilize the structures concerning 2c-2e$^-$ (two-center, two-electron) and 3c-2e$^-$ bonds [3]. B clusters have been studied extensively for decades because of their potential use in drug design and catalysis [22]. Some of the clusters exhibit planar structures, while others are considered precursors for the formation of periodic 2D flat B sheets.

2.2 2D Flat Boron Sheets

2D flat B nanosheets were prepared from boron clusters (B_{80}) by transforming the clusters into planer sheets. Szwacki et al. [23] proposed a B_{80} structure (in 2007) similar to the well-known carbon cluster C_{60}. When the B_{80} cluster was unfolded in one plane, it transformed into the famous B α-sheet [24–26]. The concept of B_{80} contributes to bridging the research gap between 2D flat B and clusters. Compared

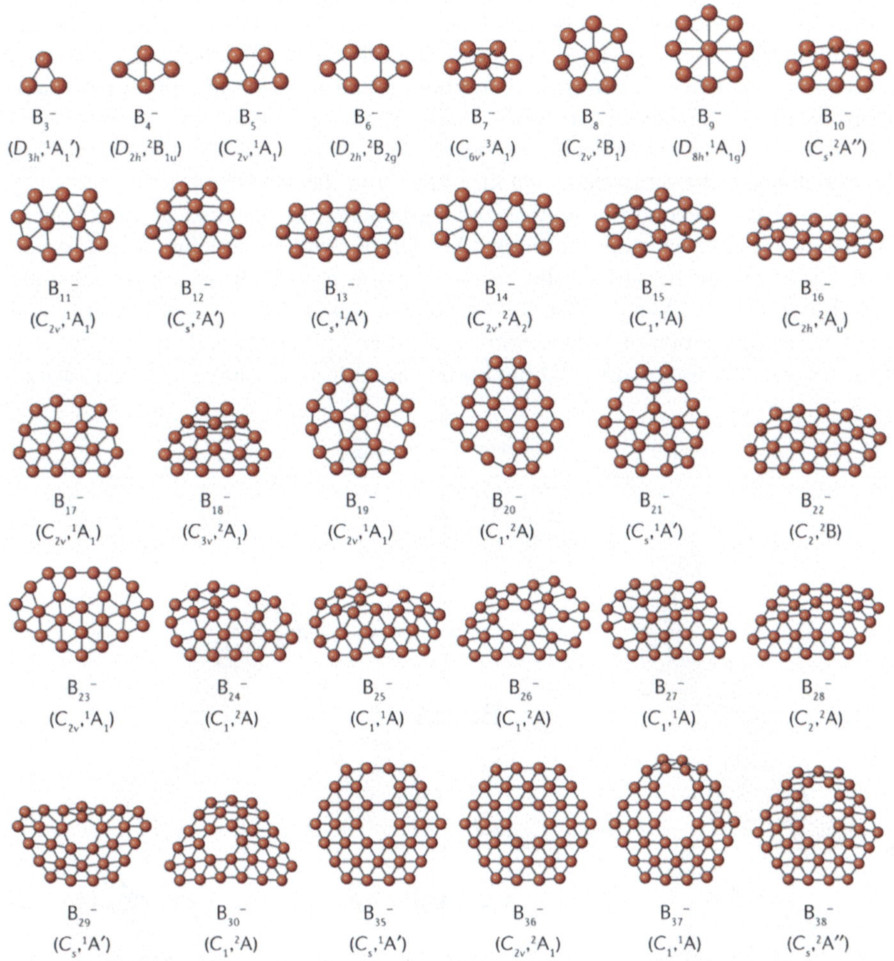

Fig. 2 Structures of B clusters of different sizes. Clusters of B_n (n = 3–30 and 35–38). Reused with permissions from [20], © Nature 2017

to Gr nanosheets, B has a honeycomb structure with sp^2 hybridized carbon, one out-of-plane π bond, and three strong in-plane σ covalent bonds, whereas B lacks one electron. The boron honeycomb becomes unstable since no electrons are occupied in the π bond [27].

To compensate for the electron deficiency in the honeycomb structure of B, the bonding mechanism of boron is modified. Generally, foreign material, such as metal layers or metal substrates, is utilized to supply the missing electrons [3]. Figure 3 illustrates the proposed structures of stabilized boron sheets, such as boron honeycomb with characteristic two-center linkages joined with B triangle motifs (a chain-like biological molecule, such as a protein or nucleic acid) [28] that serve as the fundamental building blocks for the 2D B sheet. The B honeycomb deficits electrons, but the B motif has extra electrons, and combining two forms in an appropriate ratio stabilizes the boron sheet [29]. Additionally, buckling or multilayer formation of B nanosheets can attain lower energy levels [30]. Metals may also function as an electron donor and a substrate to stabilize 2D boron sheets [31].

Li et al. have successfully synthesized a graphene-like structure of a flat honeycomb boron monolayer using aluminum (Al). The honeycomb structure was observed to be stable on the Al surface due to the electron transfer from the Al substrate to each boron atom. Metal substrates serve as a source of electronic stability and facilitate the deposition of boron atoms or clusters [32]. Zhao and co-workers also proposed a model for creating 2D flat boron sheets from clusters on the Cu surface

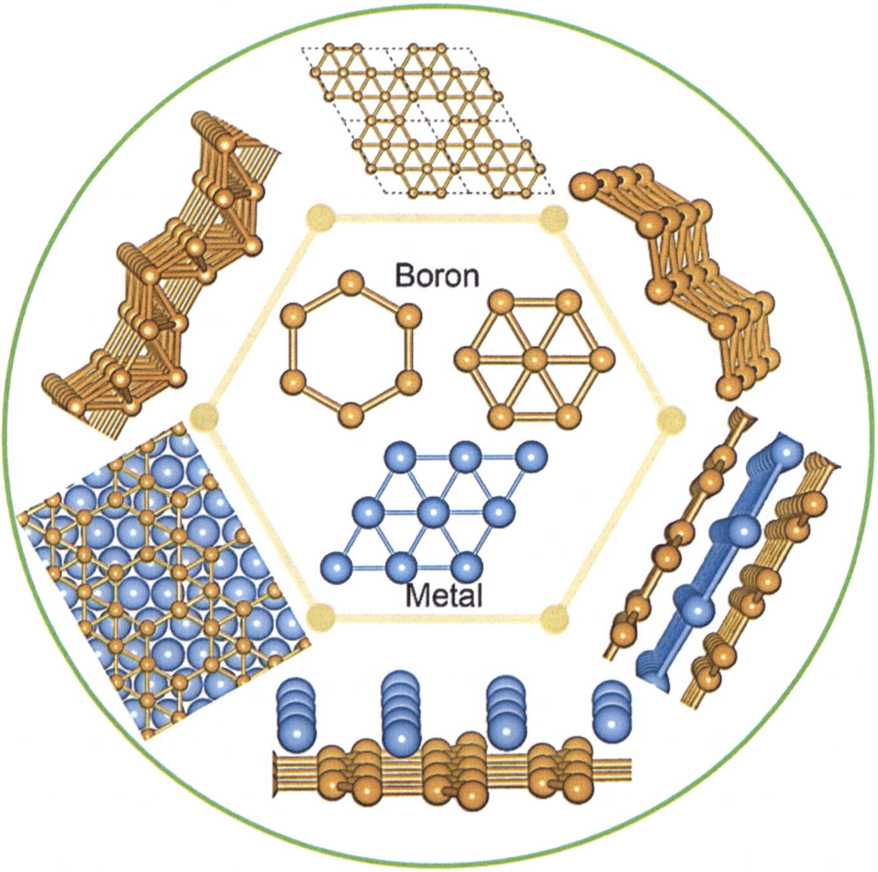

Fig. 3 Theoretical display of 2D boron sheet stabilization utilizing two-center bonds and boron motifs, buckling or formation of multilayers, metals as electron donor and substrate. Reused with permissions from [34], © Wiley 2019

and observed that when deposited boron cluster sizes increase, the clusters' energy production decreases monotonically. This was followed by a low diffusion barrier for a boron atom on Cu to ensure the growth of planar clusters. Moreover, to facilitate the large-scale production of borophenes, the required hexagonal holes may be developed at the boundaries of planar boron clusters and subsequently diffuse into the lattice [33].

Zhang et al. systematically studied the structures of 2D flat boron on Au, Ag, Cu, and Ni substrates using the cluster expansion and the surface structure search method. The results suggest that the metal substrates could lift the polymorphic energy degeneracy of isolated B to form favorable structures [31].

Unlike many emergent 2D materials, 2D flat B sheets are expected to exhibit various structures characterized by anisotropy and polymorphism that can provide enhanced property and tunability within the family of chemically similar systems. Moreover, 2D flat boron sheets can serve as a prototype for predicting and discovering new synthetic 2D materials [35, 36] in contrast to materials like Gr and silicene or semiconductors like phosphorene. The complex features of 2D structured materials, including high mechanical compliance and optical transparency, are hardly found in conventional or synthetic 2D materials. With a transition temperature range of ~10–20 K, 2D flat boron is expected to exhibit relatively high-temperature superconductivity. A novel class of 2D flat BNSs has also been investigated to broaden the scope of the 2D materials.

2.3 Borophene: Emerging 2D Boron Nanosheets

Borophene, or 2D BNSs, is an electron-deficient 2D material and is quite distinct from Gr-like structures. It has attracted extensive attention from scientific research in recent years, and it is one of the most interesting 2D nanomaterials, with a unique polymorphism behavior compared to other mono-elemental 2D materials. Borophene is similar to black phosphorus and Gr in terms of high specific surface area and drug loading capacity (borophene: 114%, [37] graphene oxide: 200%, [38] black phosphorus: 108%, [39] other nanomaterials: 10–30%) [40]. Figure 4 illustrates various 2D BNSs arrangements, emphasizing intra- and inter-planar diameters and nanosheet edges. Nano ribbons are commonly found in zigzag (B- or N-edged) or armchair (BN pair edge) conformation, with lateral diameters ranging from a few hundred nanometres to tens of microns, depending on the synthesis process. BNSs may also stack on top of one other, creating few- and multi-layers through van der Waals forces at a distance of 0.333 nm. Fullerenes, Nanotubes, and quantum dots are examples of different prevalent structural types [41].

In recent years, Certain 2D BNSs with multi-layered structures have been reported, considerably expanding the usage of 2D boron crystals [42]. Figure 4 depicts two 2D boron structures with a few atomic layer thicknesses (Pmmm-type). Evolutionary structure search techniques were used to anticipate these structures. Pmmm spacing group structures with a distorted Dirac cone at the Fermi level can be metallic or

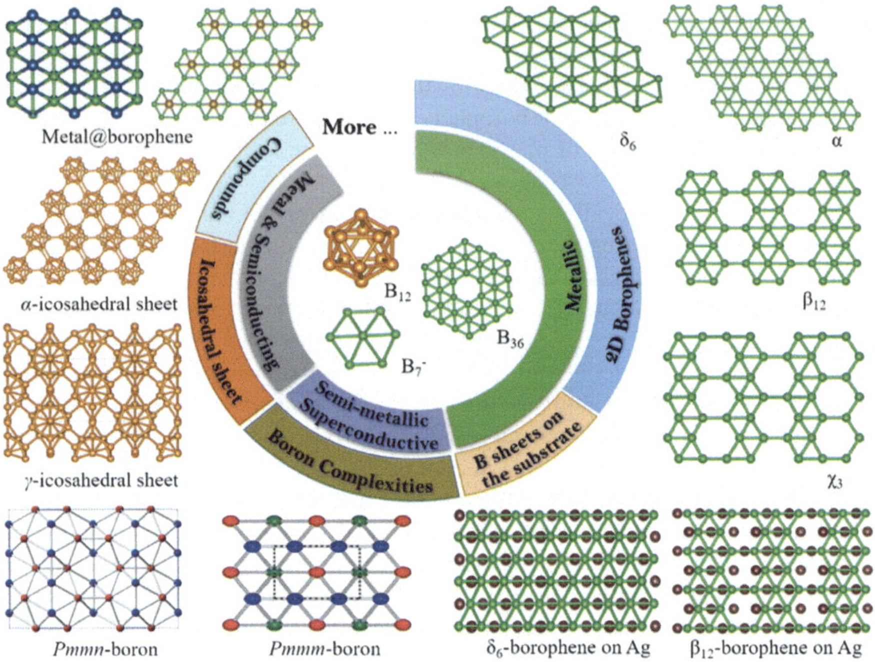

Fig. 4 Different structures of 2D BNSs. Reused with permissions from [43], © Adv. Functional material, 2017

semi-metallic. The two sublattices of B atoms, corresponding to hybrids of in-plane and out-of-plane states, were spatially inverted to generate the Dirac cone.

On the other hand, the Dirac cones in Gr were only produced from the in-plane state indicated by the π and π* bands. It has been shown that there is an alternate superconducting 2D boron structure with symmetric groups 8-Pmmm and 8-C2/m. It was stated that the electronic states of the Fermi surface with both σ and π electrons produced the superconductivity of the 2D boron structures [43].

BNSs-based composite is one of the essential classes of materials with a wide range of applications spanning from automotive, aerospace, healthcare and medicine, energy storage to electronic engineering. Boron nanomaterials have superior properties such as high Young's modulus, high thermal stability, excellent thermal expansion coefficient, and outstanding thermal conductivity. Unlike Gr and MXene, BNSs are electrically insulated due to their broad bandgap (~5.6 eV) with a dielectric constant (4 ~ 8), high voltage breakdown strength (12 $MVcm^{-1}$), and low leakage current. These characteristics imply that BNSs-based nanocomposite can be designed to achieve the desired materials with superior performances [44].

3 Properties: Physical, Chemical and Mechanical Properties

2D boron is a conventional material becoming increasingly popular as a carbon equivalent in the nano-era. The standard novel properties of 2D BNSs, such as the consistent wide bandgap, excellent anti-oxidation ability, structural stability, high thermal conductivity, and mechanical properties, provide an edge to this fascinating class of 2D materials over other materials [45]. These nanosheets have unique properties and have shown relatively better applications than carbon materials. For instance, in contrast to the black color of carbon-polymer composites, boron-polymer composites can be white or even translucent. Additionally, carbon nanostructures may ruin the electrical insulation, while boron-based nanostructures can maintain the insulation and enhance their dielectric solid characteristics [46]. Table 1 summarises various physical, chemical, and mechanical properties of 2D B nanosheets and other 2D materials.

3.1 Physical Properties

Boron is a fascinating element that is brittle (broken easily) and rare in its elemental form on Earth but frequently concentrated in the form of compounds, such as borate rocks [43]. It has a large bulk modulus in the range of (185–227 GPa) [77], a high melting point (2000 °C), and excellent rigidity [78]. These nanosheets are layered materials having a sheet-like planer structure and regularly stacked planar networks. The layered structure exhibits high flexibility to produce a wide range of nanostructures, including boron nanotubes (BNTs), BNSs, boron nano-horns (BNHs), and boron nanoparticles (BNPs). BNSs usually have a white appearance with fewer impurities (boron powder). BNSs have a large specific surface area and perform well in adsorbing organic solvents, oils, and dyes in wastewater.

3.2 Electronic and Transport Properties

Electron deficiency is one of the most essential properties of elemental boron. Boron has four accessible orbitals to be occupied, but only three valence electrons are available [79]. In addition to two-center, two-electron bonds (2c, 2e), B may also form unique multi-center, two-electron bonds because of its electron shortage. Multi-center bonding has identified up to 16 distinct B bulk allotropes. Four of these, as shown in Fig. 5, are thermodynamically stable:- α-rhombohedral, β-rhombohedral [80], γ-orthorhombic [81], and β-tetragonal boron crystals [82]. These materials are known as boron-icosahedral cluster solids (B-ICSs) [83, 84]. Boron allotropes and B-rich compounds, including BN [85], $B_{12}As_2$, $B_{12}P_2$, $B_{12}O_2$ [86], YB_{66} [87], and

Table 1 Comparison of physical, chemical and mechanical properties of 2D BNSs and other 2D nanomaterials, graphene and TMDs

S. No	Properties	BNSs	Graphene	TMDs	Refs.
1.	Color	White	Black	Shiny, dark solid	[46]
2.	Surface area (m^2/g)	~2600	~2630	~16.334	[47]
3.	Hydrophobicity	Partial hydrophilic	Both hydrophilic and hydrophobic	High hydrophobicity	[48]
4.	Cytotoxicity	Low cytotoxic	Very less cytotoxic	Lower	[49]
5.	Solubility in water	Soluble	Insoluble	Insoluble	[50]
6.	Density	2.34~2.52 g/cm^3	2.267 g/cm^3	5.06 g/cm^3	[51, 52]
7.	Melting Point	2000 °C	~4000 °C	2,375 °C	[53, 54]
8.	Chemical bonding	Covalent	Covalent	Covalent	
9.	Bond length	1.44 A°	1.42 A°	2.42 A°	[55]
10.	Bandgap (eV)	~5.6 eV	Zero	1.67 eV	[55]
11.	Cohesive energy	17.65	20.08	5.07	[56, 57]
12.	Crystallographic Parameters (nm)	a = 0.25 c = 0.666 d = 0.333	a = 0.246 c = 0.670 d = 0.335	a = 0.315 c = 1.229 d = 0.614	[58, 59]
13.	Surface energy (mJ/m^2)	35–37 (in polar solvent)	83 (in water), 115 (in N_2)	46.5 (in water)	[60]
14.	Young's modulus	0.8 TPa	1.0 TPa	0.33 ± 0.07 TPa	[61, 62]
15.	Potential barrier	Yes	No	–	[63]
16.	Galvanic corrosion	No	Yes	–	[63]
17.	Fracture strength	165 GPa	130 GPa	23 GPa	[60, 64]
18.	Thermal conductivity at room temperature	300–2000 $Wm^{-1} K^{-1}$	1800–5400 $Wm^{-1} K^{-1}$	34.5 $Wm^{-1} K^{-1}$	[45, 65]
19.	Thermal stability in air and vacuum	800 °C in the air, 1400 °C in vacuum	500–600 °C	usually 450–550 °C	[66–68]
20.	Thermal expansion coefficient ($10^{-6} K^{-1}$)	−2.72 (axial direction)	−0.8 ± 0.7	10–30	[69, 70]
21.	Dielectric constant	4 ~ 8	9.3203	5.34	[71, 72]
22.	Luminescence	Up to DUV region	No	No	[73, 74]
23.	Work function (eV)	3.65	4.49	5.05	[75, 76]

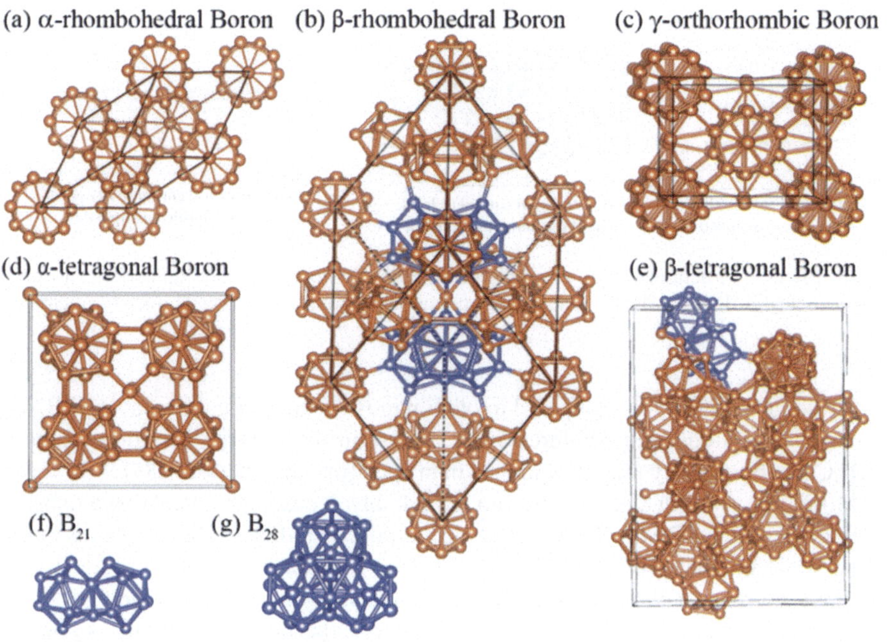

Fig. 5 Structures of stable allotropes of boron. Reproduced with permission from [89]. © Wiley 2009

boron carbide [88], can be formed by linking the icosahedral clusters with one other or with other clusters.

2D BNSs provide suitable substrates for electronic and optical applications compared to other 2D materials like Gr and MoS_2 due to their large bandgap, flat surface, and lack of surface charges. Additionally, the atomic thickness of 2D BNSs makes it possible for them to create heterostructures with other 2D nanomaterials [90]. 2D boron and their derivatives can exhibit intrinsic electrical and insulating properties due to their wide bandgap of ~5.8 eV. The inherent electrically insulating property is maintained well in BNSs as long as there are no high-concentration defects, doping, or severe edge effects owing to the thin ribbon shape. BNSs nanostructures are appropriate for polymer-based composites with adjustable dielectric characteristics [46].

3.3 Thermal Properties

Thermal properties (thermal stability and thermal conductivity) are crucial for the functioning of 2D materials. Boron nanostructures are more suitable to function under oxidation-prone conditions due to their high-performance oxidation-resistant coatings [91]. Although boron nanostructures have superior anti-oxidation properties

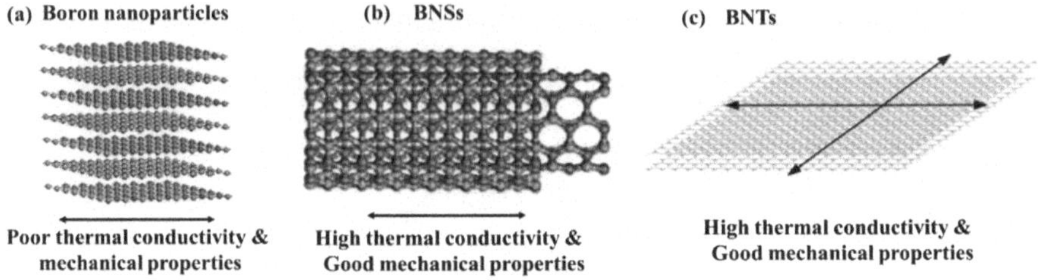

Fig. 6 Effect of structural variations on the performance of boron and their derivatives. **a** Boron nanoparticles, **b** BNSs, **c** BNTs. Reused with permissions from [46], © Royal Society Chemistry 2012

and structural stability as compared to other nanostructures, the monolayer of BNSs is unstable at higher temperatures and can sustain the thermal variation of up to 850 °C as observed by the AFM and Raman spectroscopy analyses [85].

Moreover, a strong electric field can remarkably decrease the breakdown temperature of individual BNSs. They can be decomposed at 1000–1700 °C owing to the partly ionic character of B–N bonds [92]. Due to intensive phonon conductivity, the high thermal conductivity of electrically insulating boron nanostructure is also an essential property of these materials. However, thermal conductivity is considerably lower for BNSs compared to C-based materials. For instance, the lattice thermal conductivity of BNSs is about 400 WmK^{-1} at ambient temperature, which is comparable to metals but far lower than Gr (2500 WmK^{-1}) [93]. Even though, some studies have reported thermal conductivity of 1700–3000 WmK^{-1} for BNSs using non-equilibrium Green's functions method. The strong phonon scattering constraint is primarily ascribed to the thermal conductivity of BNSs in traditional molecular dynamics methods [94].

Nanostructured boron prepared in recent years has gained unique appeal because of its high aspect ratio, mechanical stability, and superior thermal conductivity compared to the typical micro-sized boron particle fillers. The highly efficient, maximally exposed (002) lattice plane provides an extra advantage to nano-sized over conventional boron, as shown in Fig. 6. Boron particles had an average thermal conductivity of around 10 WmK^{-1}. The lattice plane exposure is maximized when the BNSs are used, enabling the operation of high in-plane thermal conductivity [46]. Therefore, single-layered BNSs are promising alternatives to electrically insulating and highly thermally conductive polymer composites.

3.4 Mechanical Properties

The mechanical strength of atomically thin BNSs has been considered superior among the class of 2D materials [95]. Boron nanostructures avoid exfoliation of boron particles, which allows high tensile strength along the plane of BNSs and

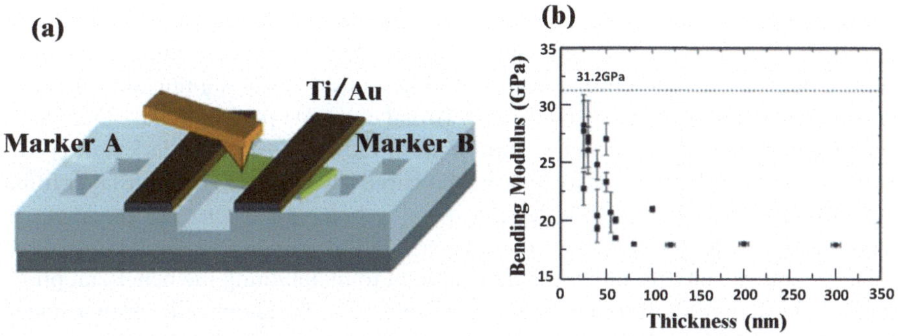

Fig. 7 **a** Schematic view to illustrate the bending of clamped BNSs, **b** Bending modulus plotted against the thickness of BNSs. Reused with permissions from [98], © IOPScience 2009

possess a high Young's modulus of 0.837 to 0.912 TPa, slightly smaller than carbon nanotubes (CNTs) [96]. Furthermore, they have greater formation energies than other 2D materials and can become more thermo-mechanically stable at high temperatures or in highly deformation durations [97]. Several theoretical models are presented to examine the mechanical properties of BNSs. The bending modulus of BNSs formed by the exfoliation process was studied experimentally and observed when the thickness of the sheet was less than 50 nm, as shown in Fig. 7. The bending moduli increased and eventually reached a value of 31.2 GPa. The strong thickness-dependent bending modulus was caused by the layer distribution of stacking faults [98].

Compared to other 2D materials, the mechanical properties of the B allotropes are superior. Both α-B_{12} (207~213 GPa) and β-B_{106} (185~210 GPa) have bulk moduli that are nearly twice as high as silicon. As a result, with hardness ratings of 42 GPa and 45 GPa, respectively, the materials designated as α-B_{12} and β-B_{106} are exceedingly hard with a bulk modulus of 221–241 GPa and a shear modulus of 236 GPa. However, γ-B_{28} has a higher modulus than α-B_{12} and β-B_{106}. Due to their extreme hardness, these boron allotropes have been used in cutting and abrasive tools [99]. Boron nanostructures can be used to strengthen polymers mechanically. For mechanical reinforcement in polymer composites, thin BNSs with low defect concentration and stacking faults are considered the best choice.

3.5 Optical and Chemical Properties

2D borons or their derivatives show interesting optical and chemical properties besides unique mechanical and electronic properties. The luminescence and UV–vis absorption characteristics of 2D boron, particularly the BNSs, are similar [100]. BNSs are a fantastic substitute for making visible light transparent polymer films that filter ultraviolet rays because of their optical characteristics. Additionally, BNSs incorporated into the polymer matrix may cause composite materials to emit UV light. BNTs can produce UV light when stimulated by light (photoluminescence) or electron beams (cathodoluminescence). Purification of BNTs significantly impacts the wavelengths of light that are released. Broad emission peaks, on the other hand, may be attributed to defect centers, impurities, or radiative excitonic dark states [46]. In contrast to Gr, which is transparent in the visible spectrum, boron multilayer films may shift from transparent to reflecting metallic behavior when their thickness is increased from 1 to 20/30 nm (for monolayer) [101].

The chemical stability of a material is crucial to maintaining the beneficial properties of the materials throughout its use. BNSs can be chemically grown under ultrahigh-vacuum conditions to avoid environmental degradation and contamination. However, they can still be contaminated after long-term exposure to air due to the electron deficiency of B atoms and the relatively high chemical activity of the surface. The boron nanosheets can oxidize due to the spontaneous adsorption of oxygen molecules from the air, which dissociates after crossing the energy barrier of ~0.35 eV. However, because of the strong B–O interaction and low mobility at room temperature, the dissociated O is difficult to move on the surface, eventually resulting in boron oxides [102].

3.6 *Other Properties*

Additional characteristics of 2D boron and its derivatives include the layered boron's superconductivity, biocompatibility, and wetting ability. The contact between liquid and solid surfaces caused by intermolecular interactions at the interface is called wetting [103]. Boron nanostructures exhibit poorer wetting qualities than bulk BN surfaces. Boron nanostructures, such as BNSs, mostly rely on surface roughness and shape to adjust their wetting rates. It is possible to create BNSs coatings with varying degrees of water repellence, ranging from superhydrophobicity with contact angles greater than 150° to partial hydrophilicity with contact angles of 50°. Water-repellent, self-cleaning, and anti-corrosion systems are some of the industrial uses of boron nanostructure films [104]. Boron nanostructures, particularly BNSs, have also been used in the biomedical field. However, their biocompatibility issues need to be investigated further to utilize the full potential of this fascinating class of materials. Prominent work on the biocompatibility of boron nanostructures has been carried out for BNTs. To facilitate BNTs' uniform dispersion in the cell culture media, polyethyleneimine (PEI) was used to prepare PEI-BNTs. When exposed to 5.0 mg mL^{-1}, the cells continue to show good viability, indicating the strong biocompatibility and biological applicability of PEI-BNTs.

Furthermore, it was shown that BNTs had a higher capacity for loading DNA and had excellent water solubility. Ciofani et al. [105] performed the first preliminary in vivo toxicity estimation of BNTs. The BNTs were shown to be a potentially useful material for biological and therapeutic applications after intravenous administration of BNTs coated with glycol chitosan (GC).

Boron compounds and its elemental form exhibit the superconductivity as well. Eremets and co-workers [106] initially found it in the boron compound MgB_2 in 2001. They discovered that at a transition temperature (Tc) of 6 K and under 160 GPa, β-rhombohedral boron transformed into a superconductor. Honeycomb-like boron layers were the primary cause of the superconductivity. Boron and compounds have received much attention in searching for new superconductors. The results of further studies demonstrated that the pressure-induced amorphization of boron occurs at about 100 GPa [107]. According to specific theoretical investigations, the face-centered cubic (fcc) phase was the foundation for superconductivity [108]. Other research, however, suggested that the α-Ga phase would be a strong contender for the high-pressure metallic phase. Under the pressure of 160 GPa, the superconducting transition was also seen in the α-rhombohedral B structure without any structural phase transitions. Additionally, it has been observed that Li-doped α-rhombohedral boron also exhibits superconducting properties.

4 Applications of 2D Boron Nanosheets

4.1 Energy Applications

The ability of a material to be employed as an electrode is mainly determined by its higher capacitance and strong electrical/ionic conductivity [109]. The excellent surface exposure and properties of 2D BNSs make it possible to be considered materials with extremely high energy storage capacities. Additionally, 2D BNSs possess metallic band configurations that can be utilized as electrodes for batteries or other energy storage systems since they facilitate electric conduction.

Most rechargeable batteries in the present scenario use lithium-ion (Li-ion) technology. Everyday items, such as electric automobiles and mobile phones, are powered by Li-ion batteries [110]. Although 2D BNSs interact well with Li-ions as an electrode material because of their high surface activity, the initial intercalation voltage (1.12 eV) decreases when the Li adsorption range widens [111]. Scientific research has shown that typical phosphorene, silicene, graphite, and other materials are not as effective as the lithiated phase of 2-Pmmn as an electrode material. The relative migration potential barrier for Li-ions in 2D BNSs is much inferior, just about 2.6 meV, than for silicene, phosphorene, graphite, and $Li_4Ti_5O_{12}$ [112]. The borophene's 325 meV longitudinal potential barrier causes significant anisotropy in the migration of the Li-ions on the 2D BNSs surface. Additionally, the electronic structure of Li exhibits the same properties of metal elements throughout the process, demonstrating the superior conductivity of 2D BNSs. In conclusion, the 2D BNSs exhibit outstanding electronic and strong ionic conductivity, performing comparatively well during charging and discharging. This can pave the path for efficient energy storage devices.

4.2 Sensing Applications

Even though conventional materials, including CNTs and Gr gas sensors, are sensitive to various gases, 2D BNS gas sensors have recently attracted much interest for their broad prospective applications. However, they cannot detect toxic compounds like formaldehyde (HCOH). Based on the DFT analyses, researchers have considered the possible application of 2D BNSs as HCOH sensors. Calculation and logical analysis reveal that changes in borophene's conductivity allowed for significant detection of an HCOH molecule [113]. As more formaldehyde molecules are absorbed by borophene, electrical signals get stronger, indicating borophene's comparably high sensitivity to HCOH gas. In addition, borophene has a high surface-to-volume ratio and may be used to detect low-concentration gas molecules [114].

2D BNSs are frequently employed as biosensors to enhance functionality and biological applications without causing harm. These materials have drawn a lot of interest in biomedicine regarding their excellent flexibility and molecular, physical, and mechanical properties. 2D BNSs could be applied to biological DNA sequencing technology. Adenine (A), cytosine (C), guanine (G), and thymine (T), the four bases of biological DNA, are examined to determine the impact of 2D borophene's energy and electron sensitivity. It was discovered that each base produced a different conductivity, producing a separate electrical signal. The base sequence determination has been used in various ways to achieve DNA sequencing, significantly impacting how humans' biological code is decoded [115]. The use of functionalized B nanosheets in this technology will substantially accelerate the study of biology and medicine.

4.3 Other Applications

Photoacoustic (PA) imaging is a biophotonic radiation diagnosis technique that is highly promising due to its benefits in deep inspection, 3D shadow, depth resolution, image clarity, stereo resolution, and sensitive dexterity. The PEG-functionalized 2D BNSs (B-PEG NSs) are reported as potential PA imaging and photothermal agents for cancer photothermal therapy. The B-PEG NSs may also effectively load imaging agents like cyanine5.5 (Cy5.5) and chemotherapy drugs like doxorubicin (DOX) [116]. Borophene, the lightest 2D nanomaterial, has the potential to operate as a catalyst for CO_2 electro-reduction as well as the hydrogen evolution processes (HER), oxygen evolution reactions (OER), and oxygen reduction reactions (ORR). Several DFT calculations revealed that an electrocatalyst for ORR and OER may be effectively made from a transition metal atom supported by borophene. This type of electrocatalyst can be utilized effectively in renewable energy systems [117]. Properties of borophene, such as the capacity for absorption, hybridization, and optical modulation, as well as their superior optical nonlinearity and atomic layer thickness, play a crucial role in photonic and electronic devices. When assessed using Z-scan, nonlinear optical properties revealed that it could work well as a broadband optical switch for ultrafast mode-locking in mid-infrared and near-infrared laser systems. According to the results, borophene could also be used as a saturable absorber (SA) in passively mode-locked lasers and other upcoming electro-optical applications [118].

5 Author's Perspective

Although borophene's development has been moving forward quickly, its enormous potential has not yet been fully utilized. The practical applications, expanding theoretical predictions, and experimental research involving the large-scale synthesis of 2D BNSs are in high demand. However, the same conventional methods that were used to create Gr may also be used to produce various phases of borophene, producing a product with superior qualities to those of Gr and its derivatives. Therefore, the theoretical simulations have great significance in advancing our understanding of 2D boron, as they can reveal information regarding stable BNSs topologies and methods for actual experimental synthesis and help us comprehend the specific electrical and mechanical properties. Borophene is anticipated to have many intriguing uses in photothermal treatment, drug delivery, cancer multimodal imaging, and other biomedical applications.

6 Summary

It is summarized that 2D BNSs are fascinating nanomaterials with varied applications. The cohesive energies of borophene and nonplanar 2D boron sheets are lower than those of Gr, silica, and germanene for single elemental 2D materials. Thus, 2D boron crystals are more stable than arsenene, antimonene, and phosphorene but less stable than silicene, Gr, and germanene. The boron derivatives, such as 2D BNSs, enhance the boron nanostructures' performance. 2D boron is a highly thermally conductive and electrically insulating material. Compared to other materials 2D BNSs have shown high flexibility and stiffness-to-weight ratio. 2D BNSs are also a good conductor of electricity and heat and can be used for superconducts. The 2D BNSs have demonstrated electrically insulating behavior across various temperatures and humidity, making them a trustworthy option with tuneable properties. They can be used as electrolyte membranes in numerous electrochemical energy conversion and storage systems. Given the structural parameters and properties (physical, thermal, mechanical, and optical), it can be concluded that nanostructured 2D boron may be considered promising materials for diversified applications such as water treatment, electric circuits, resistant coatings, glass, and ceramic industry.

Acknowledgements **Deepak Kumar** acknowledges the CSIR-Junior Research Fellowship (**CSIR-JRF**) provided for PhD dissertation work. **Pashupati Pratap Neelratan** acknowledges the UGC-Junior Research Fellowship (**UGC-JRF**) to perform PhD dissertation work. **Anshika Gupta**

is thankful to DST, Government of India, for the **INSPIRE** fellowship for Master thesis dissertation work. **Sanjeev K. Sharma** concedes Ch. Charan Singh University, Meerut, for sanctioning the project (Dev./1043/29.06.2022/41) for research and development.

References

1. Novoselov, K.S., et al.: Electric field effect in atomically thin carbon films. Science **306**(5696), 666–669 (2004)
2. Zhang, X., et al.: Regulation of 2D graphene materials for electrocatalysis **15**(15), 2271–2281 (2020).
3. Shang, J., et al.: Two dimensional boron nanosheets: synthesis, properties and applications. Phys. Chem. Chem. Phys. **20**(46), 28964–28978 (2018)
4. García-Miranda Ferrari, A., Rowley-Neale, S. J., Banks, C.E.: Recent advances in 2D hexagonal boron nitride (2D-hBN) applied as the basis of electrochemical sensing platforms. Anal. Bioanal. Chem. **413**, 663–672 (2021)
5. Keren, R.: Boron. In: Methods of Soil Analysis: Part 3 Chemical Methods, vol. **5**, pp. 603–626 (1996)
6. Sun, Z., Martinez, A., Wang, F.: Optical modulators with 2D layered materials. Nat. Photonics **10**(4), 227–238 (2016)
7. Demirel, M.C., Vural, M., Terrones, M.: Composites of proteins and 2D nanomaterials. Adv. Func. Mater. **28**(27), 1704990 (2018)
8. Rafiq, M., et al.: Recent advances in structural engineering of 2D hexagonal boron nitride electrocatalysts. Nano Energy **91**, 106661 (2022)
9. Rasul, M.G., et al.: 2D boron nitride nanosheets for polymer composite materials. npj 2D Mater. Appl. **5**(1), 56 (2021)
10. Wang, N., et al.: A universal method for large-yield and high-concentration exfoliation of two-dimensional hexagonal boron nitride nanosheets. Mater. Today **27**, 33–42 (2019)
11. Vatanpour, V., et al.: A comprehensive review on the applications of boron nitride nanomaterials in membrane fabrication and modification. Ind. Eng. Chem. Res. **60**(37), 13391–13424 (2021)
12. Kaur, J., et al.: Bidirectional superionic conduction in surface-engineered 2D hexagonal boron nitrides. ACS Appl. Mater. Interfaces **13**(5), 6532–6544 (2021)
13. Dash, B.P., et al.: Polyhedral boron clusters in materials science. New J. Chem. **35**(10), 1955–1972 (2011)
14. Chen, Q., et al.: Chemical bonding in electron-deficient boron oxide clusters: core boronyl groups, dual 3c–4e hypervalent bonds, and rhombic 4c–4e bonds. Phys. Chem. Chem. Phys. **16**(16), 7274–7279 (2014)
15. Wang, K., et al.: Concentric dual π aromaticity in bowl-like B 30 cluster: an all-boron analogue of corannulene. Phys. Chem. Chem. Phys. **18**(33), 23304–23311 (2016)
16. Chand, H., et al.: Scalable production of ultrathin boron nanosheets from a low-cost precursor. Adv. Mater. Interfaces **9**(23), 2200508 (2022)
17. Piazza, Z.A., et al.: Planar hexagonal B36 as a potential basis for extended single-atom layer boron sheets. Nat. Commun. **5**(1), 3113 (2014)
18. Boustani, I.: Systematic lsd investigation on cationic boron clusters: B (n 2–14). Int. J. Quantum Chem. **52**(4), 1081–1111 (1994)
19. Kiran, B., et al.: Planar-to-tubular structural transition in boron clusters: B20 as the embryo of single-walled boron nanotubes. Proc. Natl. Acad. Sci. **102**(4), 961–964 (2005)
20. Li, W.-L., et al.: From planar boron clusters to borophenes and metalloborophenes. Nat. Rev. Chem. **1**(10), 0071 (2017)

21. Zhang, Z., et al.: Gate-voltage control of borophene structure formation. Angew. Chem. **129**(48), 15623–15628 (2017)
22. Lesnikowski, Z.J.: Challenges and opportunities for the application of boron clusters in drug design. J. Med. Chem. **59**(17), 7738–7758 (2016)
23. Szwacki, N.G., Sadrzadeh, A., Yakobson, B.I.: B 80 fullerene: an ab initio prediction of geometry, stability, and electronic structure. Phys. Rev. Lett. **98**(16), 166804 (2007)
24. Tang, H., Ismail-Beigi, S.: Novel precursors for boron nanotubes: the competition of two-center and three-center bonding in boron sheets. Phys. Rev. Lett. **99**(11), 115501 (2007)
25. Yang, X., Ding, Y., Ni, J.: Ab initio prediction of stable boron sheets and boron nanotubes: structure, stability, and electronic properties. Phys. Rev. B **77**(4), 041402 (2008)
26. Penev, E.S., et al.: Unfolding the fullerene: nanotubes, graphene and poly-elemental varieties by simulations. Adv. Mater. **24**(36), 4956–4976 (2012)
27. Xu, M., et al.: Graphene-like two-dimensional materials. Chem. Rev. **113**(5), 3766–3798 (2013)
28. Johansson, M.U., et al.: Defining and searching for structural motifs using DeepView/Swiss-PdbViewer. BMC Bioinformatics **13**, 1–11 (2012)
29. Sergeeva, A.P., et al.: Understanding boron through size-selected clusters: structure, chemical bonding, and fluxionality. Acc. Chem. Res. **47**(4), 1349–1358 (2014)
30. Tang, H., Ismail-Beigi, S.: First-principles study of boron sheets and nanotubes. Phys. Rev. B **82**(11), 115412 (2010)
31. Zhang, Z., et al.: Two-dimensional boron monolayers mediated by metal substrates. Angew. Chem. **127**(44), 13214–13218 (2015)
32. Li, W., et al.: Experimental realization of honeycomb borophene. Sci. Bull. **63**(5), 282–286 (2018)
33. Liu, H., Gao, J., Zhao, J.: From boron cluster to two-dimensional boron sheet on Cu(111) surface: growth mechanism and hole formation. Sci. Rep. **3**(1), 3238 (2013)
34. Xie, S.Y., Wang, Y., Li, X.B.: Flat boron: a new cousin of graphene. Adv. Mater. **31**(36), 1900392 (2019)
35. Mannix, A.J., et al.: Borophene as a prototype for synthetic 2D materials development. Nat. Nanotechnol. **13**(6), 444–450 (2018)
36. Penev, E.S., et al.: Polymorphism of two-dimensional boron. Nano Lett. **12**(5), 2441–2445 (2012)
37. Ji, X., et al.: A Novel Top-Down Synthesis of Ultrathin 2D Boron Nanosheets for Multimodal Imaging-Guided Cancer Therapy, p. e1803031 (2018)
38. Zhang, L., et al.: Functional graphene oxide as a nanocarrier for controlled loading and targeted delivery of mixed anticancer drugs. Small **6**(4), 537–544 (2010)
39. Tao, W., et al.: Black phosphorus nanosheets as a robust delivery platform for cancer theranostics. Adv Mater **29**(1) (2017)
40. Ou, M., et al.: The emergence and evolution of borophene. Adv. Sci. **8**(12), 2001801 (2021)
41. Rastogi, P.K., et al.: Graphene–hBN non-van der Waals vertical heterostructures for four-electron oxygen reduction reaction. Phys. Chem. Chem. Phys. **21**(7), 3942–3953 (2019)
42. Sachdeva, G., et al.: First-principles study of linear and nonlinear optical properties of multi-layered borophene. Computation **9**(9), 101 (2021)
43. Sun, X., et al.: Two-dimensional boron crystals: structural stability, tunable properties, fabrications and applications. Adv. Func. Mater. **27**(19), 1603300 (2017)
44. Zhi, C., et al.: Large-scale fabrication of boron nitride nanosheets and their utilization in polymeric composites with improved thermal and mechanical properties. Adv. Mater. **21**(28), 2889–2893 (2009)
45. Weng, Q., et al.: Functionalized hexagonal boron nitride nanomaterials: emerging properties and applications. Chem. Soc. Rev. **45**(14), 3989–4012 (2016)
46. Meng, W., et al.: Polymer composites of boron nitride nanotubes and nanosheets. J. Mater. Chem. C **2**(47), 10049–10061 (2014)
47. Ganesha, H., et al.: 2-Dimensional layered molybdenum disulfide nanosheets and CTAB-assisted molybdenum disulfide nanoflower for high performance supercapacitor application. Nanoscale Adv. **4**(2), 521–531 (2022)

48. Fang, L.J., et al.: Hydrophobic two-dimensional MoS2 nanosheets embedded in a polyether copolymer block amide (PEBA) membrane for recovering pyridine from a dilute solution. ACS Omega **6**(4), 2675–2685 (2021)
49. Teo, W.Z., et al.: Cytotoxicity of exfoliated transition-metal dichalcogenides (MoS2, WS2, and WSe2) is lower than that of graphene and its analogues. Chemistry **20**(31), 9627–9632 (2014)
50. Lin, Y., Williams, T.V., Connell, J.W.: Soluble, exfoliated hexagonal boron nitride nanosheets. J. Phys. Chem. Lett. **1**(1), 277–283 (2010)
51. Slack, G.A., et al.: The crystal structure and density of β-rhombohedral boron. J. Solid State Chem. **76**(1), 52–63 (1988)
52. Yu Zarechnaya, E., et al.: Synthesis of an orthorhombic high pressure boron phase. Sci. Technol. Adv. Mater. **9**(4), 044209 (2008)
53. Wentorf, R.H.: Boron: another form. Science **147**(3653), 49–50 (1965)
54. Ganz, E., et al.: The initial stages of melting of graphene between 4000 K and 6000 K. Phys. Chem. Chem. Phys. **19**(5), 3756–3762 (2017)
55. Li, L.H., et al.: Large-scale mechanical peeling of boron nitride nanosheets by low-energy ball milling. J. Mater. Chem. **21**(32), 11862–11866 (2011)
56. Şahin, H., et al.: Monolayer honeycomb structures of group-IV elements and III-V binary compounds: first-principles calculations. Phys. Rev. B **80**(15), 155453 (2009)
57. Ding, Y., et al.: First principles study of structural, vibrational and electronic properties of graphene-like MX_2 (M=Mo, Nb, W, Ta; X=S, Se, Te) monolayers. Phys. B **406**, 2254–2260 (2011)
58. Wang, J., Ma, F., Sun, M.: Graphene, hexagonal boron nitride, and their heterostructures: properties and applications. RSC Adv. **7**(27), 16801–16822 (2017)
59. Zhang, K., et al.: Two dimensional hexagonal boron nitride (2D-hBN): synthesis, properties and applications. J. Mater. Chem. C **5**(46), 11992–12022 (2017)
60. Gurarslan, A., et al.: Surface-energy-assisted perfect transfer of centimeter-scale monolayer and few-layer MoS2 films onto arbitrary substrates. ACS Nano **8**(11), 11522–11528 (2014)
61. Boldrin, L., et al.: Effective mechanical properties of hexagonal boron nitride nanosheets. Nanotechnology **22**(50), 505702 (2011)
62. Li, X., Zhu, H.: Two-dimensional MoS2: Properties, preparation, and applications. J. Mater. **1**(1), 33–44 (2015)
63. Li, L., et al.: Nanosheets: boron nitride nanosheets for metal protection (Adv. Mater. Interfaces 8/2014). Adv. Mater. Interfaces **1**, p. n/a-n/a (2014)
64. Verma, A., Parashar, A., Packirisamy, M.: Tailoring the failure morphology of 2D bicrystalline graphene oxide. J. Appl. Phys. **124**(1) (2018)
65. Zhang, C., Wang, C., Rabczuk, T.: Thermal conductivity of single-layer MoS2: a comparative study between 1H and 1T′ phases. Phys. E **103**, 294–299 (2018)
66. Li, L.H., et al.: Strong oxidation resistance of atomically thin boron nitride nanosheets. ACS Nano **8**(2), 1457–1462 (2014)
67. Mazumder, M.R.H., et al.: Boron nitride based polymer nanocomposites for heat dissipation and thermal management applications. Appl. Mater. Today **29**, 101672 (2022)
68. Li, R., et al.: Thermal stability and high-temperature photoluminescence of chemical vapor deposited MoS2 in different atmosphere. J. Vac. Sci. Technol. A **40**(5) (2022)
69. Paszkowicz, W., et al.: Lattice parameters and anisotropic thermal expansion of hexagonal boron nitride in the 10–297.5 K temperature range. Appl. Phys. A **75**(3), 431–435 (2002)
70. Lin, Z., et al.: Thermal expansion coefficient of few-layer MoS2 studied by temperature-dependent Raman spectroscopy. Sci. Rep. **11**(1), 7037 (2021)
71. Pereira, V.M., et al.: Characterization of dielectric properties of graphene and graphite using the resonant cavity in 5G test band. J. Microw. Optoelectron. Electromagn. Appl **22** (2023).
72. Hou, Y., et al.: Quantification of the dielectric constant of MoS2 and WSe2 Nanosheets by electrostatic force microscopy. Mater. Charact. **193**, 112313 (2022)
73. Yu, J., et al.: Synchrotron photoluminescence spectroscopy of boron nitride nanotubes with different metal impurities. MRS Online Proc. Libr. **1204**(1), 802 (2010)

74. Splendiani, A., et al.: Emerging photoluminescence in monolayer MoS2. Nano Lett. **10**(4), 1271–1275 (2010)
75. Kim, J.H., et al.: Work function variation of MoS2 atomic layers grown with chemical vapor deposition: The effects of thickness and the adsorption of water/oxygen molecules. Appl. Phys. Lett. **106**(25) (2015)
76. Auwärter, W.: Hexagonal boron nitride monolayers on metal supports: versatile templates for atoms, molecules and nanostructures. Surf. Sci. Rep. **74**(1), 1–95 (2019)
77. Nelmes, R.J., et al.: Neutron- and x-ray-diffraction measurements of the bulk modulus of boron. Phys. Rev. B Condens. Matter **47**(13), 7668–7673 (1993)
78. Solozhenko, V.L., Kurakevych, O.O., Oganov, A.R.: On the hardness of a new boron phase, orthorhombic γ-B28. J. Superhard Mater. **30**(6), 428–429 (2008)
79. Jemmis, E.D., Balakrishnarajan, M.M.: Polyhedral boranes and elemental boron: direct structural relations and diverse electronic requirements. J. Am. Chem. Soc. **123**(18), 4324–4330 (2001)
80. Sands, D.E., Hoard, J.L.: Rhombohedral elemental boron. J. Am. Chem. Soc. **79**(20), 5582–5583 (1957)
81. Oganov, A.R., et al.: The high-pressure phase of boron, γ-B28: disputes and conclusions of 5 years after discovery. J. Superhard Mater. **33**(6), 363–379 (2011)
82. Vlasse, M., et al.: Crystal structure of tetragonal boron related to α-AlB12. J. Solid State Chem. **28**(3), 289–301 (1979)
83. Albert, B., Hillebrecht, H.: Boron: elementary challenge for experimenters and theoreticians. Angew. Chem. Int. Ed. Engl. **48**(46), 8640–8668 (2009)
84. Ogitsu, T., Schwegler, E., Galli, G.: β-rhombohedral boron: at the crossroads of the chemistry of boron and the physics of frustration. Chem. Rev. **113**(5), 3425–3449 (2013)
85. Yin, J., et al.: Boron nitride nanostructures: fabrication: functionalization and applications. Small **12**(22), 2942–2968 (2016)
86. Becher, H.J., Thévenot, F.: Infrarotspektroskopische untersuchung des borcarbids und seiner isotypen derivate B12O2, B12P2 und B12As2. Z. Anorg. Allg. Chem. **410**(3), 274–286 (1974)
87. Richards, S.M., Kaspar, J.S.: The crystal structure of YB66. Acta Crystallogr. B **25**(2), 237–251 (1969)
88. Balakrishnarajan, M.M., Pancharatna, P.D., Hoffmann, R.: Structure and bonding in boron carbide: the invincibility of imperfections. New J. Chem. **31**(4), 473–485 (2007)
89. Albert, B., Hillebrecht, H.: Boron: elementary challenge for experimenters and theoreticians. Angew. Chem. Int. Ed. **48**(46), 8640–8668 (2009)
90. Dean, C.R., et al.: Boron nitride substrates for high-quality graphene electronics. Nat. Nanotechnol. **5**(10), 722–726 (2010)
91. Lin, Y., et al.: Aqueously dispersed silver nanoparticle-decorated boron nitride nanosheets for reusable, thermal oxidation-resistant surface enhanced Raman spectroscopy (SERS) devices. ACS Appl. Mater. Interfaces **4**(2), 1110–1117 (2012)
92. Turhan, E.A., et al.: Properties and applications of boron nitride nanotubes. Nanotechnology **33**(24), 242001 (2022)
93. Guerra, V., Wan, C., McNally, T.: Thermal conductivity of 2D nano-structured boron nitride (BN) and its composites with polymers. Prog. Mater Sci. **100**, 170–186 (2019)
94. Huang, J., et al.: Molecular dynamics investigation of the thermal properties in single-walled boron nitride nanotube. Mater. Res. Express **7**(2), 025025 (2020)
95. Cellini, F., et al.: Pressure-induced formation and mechanical properties of 2D diamond boron nitride. Adv. Sci. **8**(2), 2002541 (2021)
96. Arenal, R., et al.: Young modulus, mechanical and electrical properties of isolated individual and bundled single-walled boron nitride nanotubes. Nanotechnology **22**(26), 265704 (2011)
97. Kim, J.H., et al.: Boron nitride nanotubes: synthesis and applications. Nano Converg. **5**(1), 1–13 (2018)
98. Li, C., et al.: Thickness-dependent bending modulus of hexagonal boron nitride nanosheets. Nanotechnology **20**(38), 385707 (2009)

99. Lyakhov, A.O., Oganov, A.R., Valle, M.: Crystal structure prediction using evolutionary approach. Mod. Methods Cryst. Struct. Predict. 147–180 (2010)
100. Kumbhakar, P., et al.: Nonlinear optical properties and temperature-dependent uv–vis absorption and photoluminescence emission in 2d hexagonal boron nitride nanosheets. Adv. Opt. Mater. **3**(6), 828–835 (2015)
101. Adamska, L., et al.: First-principles investigation of borophene as a monolayer transparent conductor. J. Phys. Chem. C **122**(7), 4037–4045 (2018)
102. Luo, W., et al.: The adsorption and dissociation of oxygen on Ag (111) supported χ3 borophene. Phys. B **537**, 1–6 (2018)
103. Fowkes, F.M.: Attractive forces at interfaces. Ind. Eng. Chem. **56**(12), 40–52 (1964)
104. Pakdel, A., et al.: Low-dimensional boron nitride nanomaterials. Mater. Today **15**(6), 256–265 (2012)
105. Ciofani, G., et al.: Boron nitride nanotubes: Biocompatibility and potential spill-over in nanomedicine. Small **9**(9–10), 1672–1685 (2013)
106. Kortus, J., et al.: Superconductivity of metallic boron in MgB 2. Phys. Rev. Lett. **86**(20), 4656 (2001)
107. Sanz, D.N., Loubeyre, P., Mezouar, M.: Equation of state and pressure induced amorphization of β-boron from X-ray measurements up to 100 GPa. Phys. Rev. Lett. **89**(24), 245501 (2002)
108. Luo, H.: Superconductivity and lattice parameters in face-centered cubic Pt-W and Pd-W solid solutions. J. Less Common Metals **15**(3), 299–302 (1968)
109. Weingarth, D., et al.: Graphitization as a universal tool to tailor the potential-dependent capacitance of carbon supercapacitors. Adv. Energy Mater. **4**(13), 1400316 (2014)
110. Hannan, M.A., et al.: State-of-the-art and energy management system of lithium-ion batteries in electric vehicle applications: Issues and recommendations. IEEE Access **6**, 19362–19378 (2018)
111. Qiu, S., et al.: Manipulating adsorption–insertion mechanisms in nanostructured carbon materials for high-efficiency sodium ion storage. Adv. Energy Mater. **7**(17), 1700403 (2017)
112. Tian, Y., et al.: Elemental two-dimensional materials for Li/Na-ion battery anode applications. Chem. Rec. **22**(10), e202200123 (2022)
113. Kootenaei, A.S., Ansari, G.: B36 borophene as an electronic sensor for formaldehyde: quantum chemical analysis. Phys. Lett. A **380**(34), 2664–2668 (2016)
114. Opoku, F., Govender, P.P.: Highly selective and sensitive detection of formaldehyde by β12-borophene/SnO2 heterostructures: the role of an external electric field and in-plain biaxial strain. J. Phys. Chem. A **124**(11), 2288–2300 (2020)
115. Rastgou, A., Soleymanabadi, H., Bodaghi, A.: DNA sequencing by borophene nanosheet via an electronic response: a theoretical study. Microelectron. Eng. **169**, 9–15 (2017)
116. Ji, X., et al.: A novel top-down synthesis of ultrathin 2D boron nanosheets for multimodal imaging-guided cancer therapy. Adv. Mater. **30**(36), 1803031 (2018)
117. Chand, H., Kumar, A., Krishnan, V.: Borophene and boron-based nanosheets: recent advances in synthesis strategies and applications in the field of environment and energy. Adv. Mater. Interfaces **8**(15), 2100045 (2021)
118. Ma, C., et al.: Broadband nonlinear photonics in few-layer borophene. Small **17**(7), 2006891 (2021)
119. Pal, V.K., Kumar, D., Gupta, A., Neelratan, P.P., Purohit, L.P., Singh, A., Singh, V., Lee, S., Mishra, Y.K., Kaushik, A., Sharma, S.K.: Nanocarbons decorated TiO_2 as advanced nanocomposite fabric for photocatalytic degradation of methylene blue dye and ciprofloxacin. Diam. Relat. Mater. **148**, 111435 (2024). https://doi.org/10.1016/j.diamond.2024.111435
120. Upadhyay, G.K., Purohit, L.P., Sharma, H., Jain, N., Sharma, S.K.: Z-scheme based photoactive ZnO: TiO_2: CdO: g-C_3N_4 nanocomposites for advance oxidation process. J. Mol. Struct. **1319**, 139366 (2025). https://doi.org/10.1016/j.molstruc.2024.139366
121. Kumar, D., Neelratan, P.P., Gupta, A., Sharma, N., Sharma, M., Shukla, S., Singh, S.P., Yu, J.S., Kaushik, A., Sharma, S.K.: Carbon-based metal-oxides and MOFs for efficient CO_2 detection/reduction to chemical/fuels. Mater. Today Sustain. **28**, 100952(2024). https://doi.org/10.1016/j.mtsust.2024.100952

122. Tyagi, N., Sharma, G., Kumar, D., Neelratan, P.P., Sharma, D., Khanuja, M., Singh, M.K., Singh, V., Kaushik, A., Sharma, S.K. (2023) 2D-MXenes to tackle wastewater: from purification to SERS-based sensing. Coord. Chem. Rev. **496**, 215394 (2023). https://doi.org/10.1016/j.ccr.2023.215394
123. Misra, V., Singh, V., Singh, A., Kumar, D., Sharma, S.K.: Photocatalytic degradation of methylene blue dye in aqueous solution under UV irradiation using $NdMnO_3$: rGO hybrid nanocomposites as a photocatalyst. Appl. Surf. Sci. Adv. **18**, 100491 (2023). https://doi.org/10.1016/j.apsadv.2023.100491
124. Jain, N., Kumar, D., Bhardwaj, K., Sharma, R.K., Holovsky, J., Mishra, M., Mishra, Y.K., Sharma, S.K.: Heterostructured core-shell metal oxide-based nanobrushes for ultrafast UV photodetectors. Mater. Sci. Eng.: R Rep. **160**, 100826 (2024). https://doi.org/10.1016/j.mser.2024.100826

Borophene: A New Analog of Graphene

Sagar Mondal, Swati Priya, Smita Lata, Shivam Kumar, Himanshu Ranjan, Sarvesh Rustagi, Mohd. Abubakar Sadique, Shalu Yadav, Jutishna Bora, Sumira Malik, and Raju Khan

Abstract Borophene is a Dirac material made of boron atoms with its two-dimensional structure offering great electrical characteristics like graphene. The hexagonal monolayer structure of boron atoms, similar to the carbon atoms in graphene, is dependent on the atomic structure and arrangement of boron. This material has promise in electronics, energy storage, catalysis, and sensing. Scientists and academicians at a global level have become increasingly interested in its importance as a nanotechnology material in recent years. It is sometimes called a "novel counterpart to graphene" because of its unique properties and broad applications. The material has good ideal specific capacity, ion transportation, and electric conductivity. It can be used as a negative electrode in lithium-ion batteries. It also has great in-plane strength and flexibility and detects atoms and molecules due to its catalytic properties. The high hydrogen adsorption affinity of borophene is that this phenomenon catalyzes the decomposition of molecular hydrogen into hydrogen ions and reduces water. Borophene is versatile in catalysis, electronics, energy storage, and quantum computing because of its high electric conductivity, flexibility, and thermal efficiency. Borophene research expands its uses in sophisticated materials and technologies.

Keywords Borophene · Graphene · Energy storage · Catalysis

Sagar Mondal and Swati Priya: These author are equally contributed

---

S. Mondal · S. Priya · S. Lata · S. Kumar · H. Ranjan · J. Bora · S. Malik (✉)
Amity Institute of Biotechnology, Amity University, Jharkhand, Ranchi, India
e-mail: smalik@rnc.amity.edu

S. Rustagi
Department of Food Technology, Uttaranchal University, Dehradun, Uttarakhand, India

Mohd. A. Sadique · S. Yadav · R. Khan (✉)
Industrial Waste Utilization, Nano and Biomaterials, CSIR-Advanced Materials and Processes Research Institute (AMPRI), Hoshangabad Road,, Bhopal, MP 462026, India
e-mail: rajukhan.ampri@gmail.com

Academy of Scientific and Innovative Research (AcSIR), Ghaziabad 201002, India

R. Khan et al. (eds.), *2D Boron Nanosheets*, Engineering Materials,
https://doi.org/10.1007/978-981-97-7014-4_3

1 Introduction

One remarkable discovery has recently captured the interest and imagination of researchers in the field of advanced materials science. Borophene a two-dimensional material (2DM) has caught the attention of scientists because of its extraordinary qualities and potential for completely transforming many industries [1]. Borophene, which is frequently referred to as a new analog of graphene, is a cutting-edge advancement in the search for novel materials with exceptional electrical, mechanical, and thermal properties. Graphene, a substance composed of carbon atoms arranged in a hexagonal structure, gained considerable interest after its successful isolation and characterization by Andre Geim and Konstantin Novoselov in 2004 [2]. Their groundbreaking accomplishment was duly recognized with the awarding of the Nobel Prize in Physics. The exceptional electrical conductivity, remarkable thermal conductivity, and extraordinary strength exhibited by graphene have spurred extensive research in various fields. Nevertheless, scientists have continuously strived to find materials that exhibit comparable or superior properties. As a result, borophene emerged as a key discovery in this ongoing search [1]. Borophene, a recently discovered member of the Xene family, is a material composed of a single layer of boron atoms and their significant contribution leads to an in-depth understanding of the adsorption mechanisms involved in aromatic π-systems of 2D graphene and borophene sheets in the wide range of applications [2]. It is noteworthy for its remarkable electrical, superconducting, elastic, thermal, anisotropic mechanical, and optical characteristics. These attributes are well-suited for a wide range of sophisticated technological applications, covering the storage of energy, biological sensing, detection of gases, medications, catalysis, and more. As we write this chapter, we will explore the world of borophene, a new 2D material composed entirely of boron atoms, revealing its synthesis methods, various properties, potential applications, recent advances, and prospects [3]. A comparative assessment was conducted to evaluate the ability of striped borophene (sB), β_{12} borophene (β_{12}), and pristine graphene (GN) to adsorb π-systems. Benzene (BNZ) and hexachlorobenzene (HFB) were used as electron-rich and electron-deficient aromatic π-systems, respectively as shown in Fig. 1 [2].

2 Synthesis of Borophene

To synthesize borophene, the following methods have played a pivotal role, which are explained in various sections.

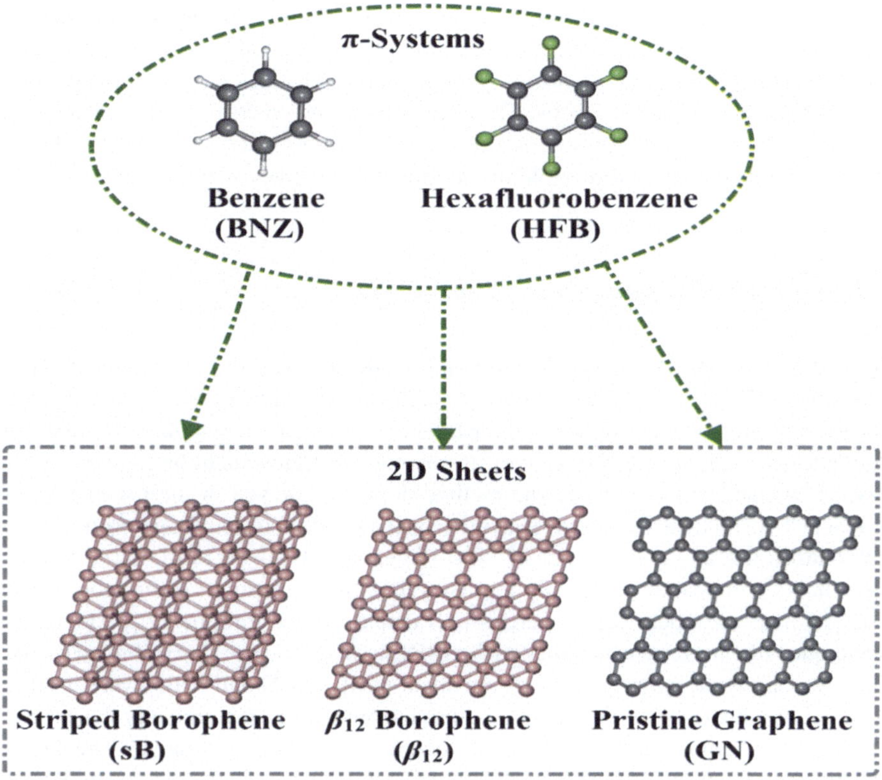

Fig. 1 Sig contribution leads to an in-depth understanding of the adsorption mechanisms involved in aromatic π-systems of 2D graphene and borophene sheets in a wide range of applications. Reprinted (adapted) with permission from [2]. Copyright 2022 MDPI

2.1 Substrate-Supported Crystal Growth Using Atomic Layer Deposition

In 2015, Mannix and the research team conducted the very first synthesis of borophene using atomic layer deposition (ALD) on the Ag (111) surface. The synthesis process involved a deposition rate ranging from 0.01 to 0.1 monolayer per minute, an acceleration voltage between 1.8 and 2.3 kilovolts, a heat flow of 1.5 to 1.9 amperes, and a pressure of 0.5×10^{10} millibar. A borophene layer was generated through the facilitation of crystallization by a silver substrate. They also stated that it showed anisotropic phases χ_3 (uniform phase) and 12 (wavy "striped" phase) of borophene at two significantly higher temperatures (450 °C and 700 °C, respectively). Stable striped facies are produced by high deposition temperatures and low deposition rates. Metastable phases are more favorable at low temperatures and fast deposition rates. The presence of structural anisotropy in the form of periodic vertical buckling, short-range rhombohedral moiré patterns, and long-range one-dimensional moiré patterns was verified through the utilization of atomic force microscopy and high-resolution transmission electron microscopy. Additionally, the phase purity and the absence or presence of impurities in compounds and alloys were confirmed using X-ray photoelectron spectroscopy. The utilization of low-energy electron diffraction techniques in the analysis of microscopic pictures has provided evidence that the χ_3 and β_{12} phases possess a three-fold orientational degeneracy with the substrate.

2.2 Chemical Vapour Deposition (CVD)

Some 2D materials can be synthesized on an industrial scale using chemical vapor deposition (CVD). The CVD technique has drawn significant interest in recent times due to its prominent position in the industry and its capacity to manufacture sheets of graphene of exceptionally large sizes on several substrates, including copper (Cu), silver (Ag), and gold (Au) like the methodologies employed in the production of graphene [4]. Tai et al. successfully synthesized a material resembling borophene on CVD substrates at a temperature of 1100 °C, using a B: B_2O_3 ratio of 1:1. On the other hand, composites are mostly multi-layered and made up of a common phase. The physical and chemical properties of vacuum-grown borophene, produced by techniques like molecular beam epitaxy (MBE), ALD, and CVD, exhibit variations that are contingent upon its level of chemical purity. ALD and MBE were used to deposit borophene sheets. By injecting diborane onto a heated atomically flat surface under high vacuum conditions, the CVD technique produces borophene. The chemical was subsequently isolated from the borazine by a freeze–thaw process employing a liquid nitrogen cold trap. In an environment of ultrahigh vacuum, atomically thin, crystalline 2D boron sheets are synthesized on a silver surface [5]. In the conventional approach, the pure boron source is subjected to direct evaporation, resulting in the production of boron. Subsequently, the boron material is cultivated on the surface of a single crystal Ag (111). Throughout the process of growth, the substrate needs to be exposed to elevated temperatures, often reaching approximately 500 K. During this procedure, the condensation of hot gas containing boron atoms occurs on the frigid surface of silver that is in its pure form. Every individual atom forms chemical bonds with six neighboring atoms, resulting in a planar hexagonal arrangement [5]. Additionally, the presence of silver atoms induces the boron atoms to adopt a similar structure. Nevertheless, the current arrangement exhibits deficiencies since numerous boron atoms form bonds with merely four or five neighboring atoms. The special properties of borophene crystals are due to the configuration of the pores.

The researchers Wu et al. have devised an enhanced technique for the CVD of micrometre-scale single-layer borophene sheets on thick Cu (111) films grown on sapphire substrates. This advancement extends the capabilities of borophene-based devices and lays the foundation for the development of related electrical devices.

2.3 Molecular Beam Epitaxy (MBE)

In 2016, Feng et al. conducted a study to validate the 2D structure of borophene. This was achieved by synthesizing atomic sheets of borophene on Ag (111) substrates. The synthesis process involved utilizing MBE at temperatures ranging from 550 to 800 K. It is worth noting that the pressure during boron growth was maintained at 6×10^{-10} Torr. A high-purity boron sample, consisting of 99.9999% boron, was subjected to evaporation onto a pristine Ag (111) substrate. The cleaning process of Ag (111) single crystals involved multiple cycles of sputtering using argon ions, accompanied by a boron flux of around 0.1 monolayers per minute. The β_{12} and χ_3 phases were discovered by Feng et al. and Mannix et al. Upon conducting a meticulous analysis of their atomic composition, it was observed that, despite the utilization of distinct deposition techniques, the boron atomic concentrations in the two phases remained discernible. Borophene growth using ALD and MBE essentially limits borophene deposition to limited areas, making it difficult and expensive to produce industrially usable devices. The advantages and disadvantages of various synthesis methods of borophene are listed in Table 1.

Table 1 Comparison of methodology of borophene synthesis-

Synthesis method	Description	Advantages	Disadvantages
Substrate supported crystal growth with ALD	ALD is employed to add boron atoms arranged in the substrate in a controlled manner	Precise control over film thickness and composition	Limited availability of suitable boron precursors. Requires complex equipment and high-purity substrates
Chemical Vapour Deposition (CVD)	Precursor gases containing boron are decomposed on a substrate at elevated temperatures	High-quality borophene films can be produced	Requires specialized equipment. Precursor availability can be limited
Molecular Beam Epitaxy (MBE)	Boron atoms are deposited onto a substrate with controlled flux and energy	Precisely controlled over film thickness and composition	Expensive and complex equipment. Requires ultra-high vacuum conditions

3 Properties of Borophene

3.1 Mechanical Properties

The mechanical characteristics of 2D materials, exhibit an ultra-thin morphology and are influenced by the manipulation of lattice stress and strain. Additionally, it is important to consider the stress of electronic devices fabricated with 2D materials Xenes. Most of the existing theoretical investigations on 2D borophene primarily rely on first-principles simulations. In this study, Zhou and Jiang employed molecular dynamics simulation (MDS) techniques to investigate the mechanical characteristics of 2D borophene [6]. The mechanical parameters that were determined exhibited a high level of consistency with the results obtained from previous ab initio calculations. The mechanical strength of freestanding borophene sheets in borophene nanocomposites including polyvinylidene difluoride (PVDF) was examined by Ranjan et al. The most favorable outcomes in terms of elongation were obtained when utilizing a self-supporting layer of borophene (a type of nanofiller) with an average concentration of 1%. Due to its B–B bond, the original borophene layer was predicted to exhibit high Young's modulus and in-plane Poisson's ratio anisotropy. The distinctive structure of boron offers superior mechanical qualities compared to conventional 2D materials. In a recent study conducted by Zhang et al., it was demonstrated by a first-principles calculation that borophene sheets, characterized by a concentration of hollow hexagons at a ratio of 1/6, had a bending stiffness that is twice as high as that of graphene [7]. Additionally, it was found that the delocalized and multicentric chemical bonds of borophene undergo a structural phase transition before being disrupted by strain. Even more than graphene, borophene films without H–H bonds are promising as advanced composite materials. The mechanical strength of the borophene layer is mainly determined by its dimensional reduction, anisotropy, various polymorphisms, and lateral size [8].

3.2 Thermal Properties

Borophene is promising as a transparent conductive and thermal management material due to its anisotropy of heat transport and low thermal conductivity. The thermal characteristics of borophene exert a substantial influence on its practical utility [9]. The stability and longevity of photovoltaic and electrical devices heavily rely on the presence of high heat conductivity [10]. Conversely, materials exhibiting low thermal conductivity are well-suited for applications involving thermal electricity and heat insulation. Morzavati et al. conducted a prior investigation on the thermal properties of five distinct borophene structures using ab initio molecular dynamics (MD) simulations. Based on the outcomes of their investigation, it was determined that the borophene layer denoted as S1 exhibited a capacity to endure temperatures of up to 750 K, beyond which it underwent destruction at 2000 K. In contrast, the borophene layer S2 exhibited partial dissolution at a temperature of 2000 K, followed by full collapse with further heating to 3000 K. Additional variants, namely S3 and S5, exhibited structural stability up to a temperature of 2000 K, but failed to retain their integrity when exposed to a higher temperature of 3000 K. The thermal stability of the S4 borophene layer was comparatively poorer than that of the S3 and S5 layers, as its structural integrity was seen to be susceptible to degradation even at temperatures as low as 2000 K. Based on the findings, it was seen that the borophene layer exhibited resilience within the temperature range of 750 to 2000 K. However, upon above this threshold, the structural integrity of the layer experienced either complete or partial collapse, contingent upon the specific atomic configuration.

Xiao et al. employed first-principles calculations and the phonon Boltzmann transport equation in this study to examine the thermal conductivity properties of warped borophene plates, with a lattice constant of $a = 4.37$ and a vertical plane distance of 0.173. Based on empirical findings, it has been observed that the heat transfer capacity of borophene exhibits an initial positive correlation with temperature, up to a threshold of 150 K. However, beyond this critical temperature, the thermal conductivity of borophene demonstrates a subsequent decline. This phenomenon occurs due to the greater phonon–phonon scattering that accompanies the increase in temperature. In addition, it was observed that the thermal conductivity of the lattice of the borophene sheet exhibited a notable reduction (14.3 W mK) in comparison to graphene (about 3500 W mK), indicating its efficacy as a thermal insulator [11]. It is noteworthy that the thermal conductivity of borophene is predominantly influenced by high-frequency phonon modes, accounting for about 70% of its overall thermal conductivity. The characteristic that sets this material apart from other 2DMs, such as graphene, silicene, and phosphorene, is the influence of low-frequency phonon modes on its transport properties.

3.3 Electron and Optical Properties

Boron, the fifth element on the periodic table, is the only semiconductor in group 13 and has valence orbitals like carbon. Thus, "Borophene" and graphene, a 2D carbon structure, have similar characteristics. Borophene has unique polymorphism compared to other 2D single-element materials as depicted in Fig. 2 [12].

Electronic Properties

The electronic characteristics of the 2-Pmmn phase of borophene exhibit significant anisotropy in its electrical structure. The structure of the electronic strip exhibits metallic conduction along the Γ-X and Y-S axes, particularly in the armchair direction. A group velocity of 6.6×10^5 m/s was achieved. Conversely, the band structure exhibits a band gap in the Γ-Y and S-X directions, which correspond to the zigzag direction [13].

The utilization of chemical modification is a highly effective approach for the manipulation of the band structure of 2D materials. The process of hydrogenation

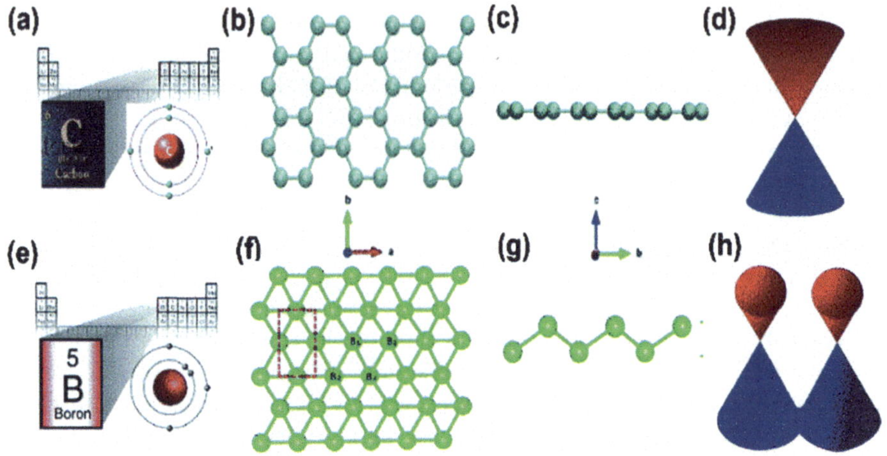

Fig. 2 The valence electrons of carbon and boron are depicted schematically in their respective comparable orbitals. Graphene is shown from a top view (**b**), side view (**c**), and its Dirac cone (**d**). Borophene is shown from a top view (**f**), side view (**g**), and its Dirac cone (**h**). Reprinted (adapted) with permission from [12]. Copyright 2022 Royal Society of Chemistry

and fluorination has been employed to induce band gap opening in graphene, silicene, and germanene. The metallic nature of the band structure in the 2-Pmmn phase of borophene has been observed. Conversely, the band structure of the 2-Pmmn phase of fully hydrogenated borophene transforms, resulting in a semiconductor with a zero energy gap [14]. The optoelectronic properties of two possible allotropes of 2D-borophene, namely β_{12} and δ_6, were investigated by Lyudmyla Adamska et al. through the use of fundamental principles of DFT and many-particle perturbation theory [15]. Both systems exhibit characteristics of anisotropic metals, including anisotropic thickness, energy-dependent optical transparency, and low absorption levels, specifically accounting for only 1% of the visible spectrum. Furthermore, the employment of novel methodologies in elucidating the interactions between electrons and phonons reveals that electron–phonon couplings have restricted conductivity. These results show that both allotropes can be used as transparent electrodes. The predominant characteristic of 2D borophene is its metallic behavior, which stands in contrast to the insulating nature of 3D bulk formations [16].

Optical Properties

The presence of an anisotropic atomic structure in borophene results in a notable optical anisotropy within the plane. The study revealed that the orientation of borophene had a notable influence on its reflectance within the visible spectrum. The transparent metallic properties and energy-dependent optical transparency of borophene polymorphs β_{12} and δ_6 have been observed, indicating substantial thickness [17]. The presence of a significant electron–phonon interaction results in a reduced absorption (1%) of visible light in both phases. In their study, Lherbier et al. undertake a comparative analysis of the optical conductivity exhibited by pure borophene and pure graphene [13]. The experimental results indicate that borophene exhibits a poorer optical conductivity (σ) compared to single-layer graphene when subjected to low-energy radiation with a wavelength as low as 3 eV, which falls within the lower range of the visible light spectrum. However, in the ultraviolet (UV) region, the optical conductivity of borophene exhibits a progressive increase up to 6.5 eV, followed by the appearance of four distinct peaks at 6.5, 8.2, 8.5, and 9.6 eV. This indicates an amplification of the optical signal. Furthermore, borophene exhibits an extremely low level of transmission and reflection throughout the infrared spectrum. Nevertheless, it is worth noting that the optical conductivity of oxidized borophene exhibits a notable increase in both the visible and infrared ranges when compared to the unoxidized counterpart. This increase occurs within a certain range of oxidation levels, specifically between 1013 cm^2 and 1014 cm^2. Despite the considerable degree of oxidation, the optical conductivity within the visible spectrum remains comparatively lower than that observed in pure graphene. In a study conducted by Lyudmila Adamuska and colleagues, the authors employed DFT simulations to demonstrate the ability to consistently manipulate the optical absorbance and electronic band structure of 2D borophene through the application of minor stresses. Both structures exhibit metallic properties when subjected to modest applied voltages of up to 6%. However, the band structure plays a crucial role in enabling substantial adjustments to the anisotropy and optical absorption of the complicated in-plane dielectric function. The study conducted by Adamska, and colleagues revealed that β_{12}^- and δ_6^- metallic 2D borophene-Xenes have anisotropic properties. These materials demonstrate optical transparency that is significantly dependent on energy and thickness, while also exhibiting low absorption in the visible spectrum (1%). The enhancement of the transition from a transparent to a reflecting metal within the visible spectrum of multilayer films is shown to occur at a thickness range of 20–30 nm for 2D graphene material, which functions as a monolayer of metal with transparency in the visible range [18]. The argument presented is persuasive.

3.4 *Superconductive Properties*

The initial investigation on the superconducting characteristics of borophene was conducted by Penev et al. They were selected as $B_{1-x}V_x$ (B: B grid, V: Vacancy) Polymorph using first-principles calculations. According to their findings, borophene has inherent superconductivity with a critical temperature (Tc) value of approximately 10–20 K, and the Tc has a positive correlation with the reduction in the number of vacancies. In addition, the researchers asserted that the introduction of a metal surface with strong adhesion could potentially reduce the Tc of borophene under certain conditions. The conclusions made by Penev et al. have been substantiated through first-principles calculations conducted by Zhao et al. [19, 20]. It has been determined that all stable phases of borophene demonstrate superconducting characteristics. The Tc of buckling triangular borophene on the Ag(111) surface was

Table 2 Comparison of properties of Borophene

Properties	Descriptions	Characteristics
Mechanical properties	Response to mechanical forces	High strength, flexibility, anisotropic
Thermal properties	Ability to conduct and store heat	High thermal conductivity, low thermal expansion
Electronic properties	The behavior of electrons in the material	Varied electronic structures, tunable bandgap
Optical properties	Interaction with light	Transparency (in some cases), varying absorption
Superconducting properties	Conducting electricity with zero resistance	Potential for superconductivity at low temperatures

determined to be 5.2 K. The Tc value obtained in this study is comparatively lower than the value reported by Penev et al., who also observed a reduction in the Tc value for the borophene/Ag(111) system. Borophene was detected compared to free bonds. Zhao et al. from another study revealed that some borophene structures exhibit superconducting properties at Tc below the temperature of liquid hydrogen (33 K) [21]. They suggested that among all pure single-element materials, the 2D-B structure could have the highest Tc without high pressure or external stress. The peculiar properties of borophene are described in Table 2.

4 Potential Applications of Borophene

Borophene exhibits great potential for several applications including energy storage, hydrogen storage, supercapacitors, gas sensors, catalysis, electronics, and biological applications. Due to its exceptional electrical conductivity and large theoretical capacity for lithium ions, it is considered a promising option for future battery electrodes. The fact that it can reversibly interact with hydrogen molecules makes it well-suited for hydrogen fuel cells. Borophene possesses a substantial surface area and exhibits rapid charge and discharge rates, rendering it well-suited for utilization in supercapacitors. Its distinctive characteristics also render it a highly promising contender for high-performance transistors. Figure 3 shows various applications of borophene.

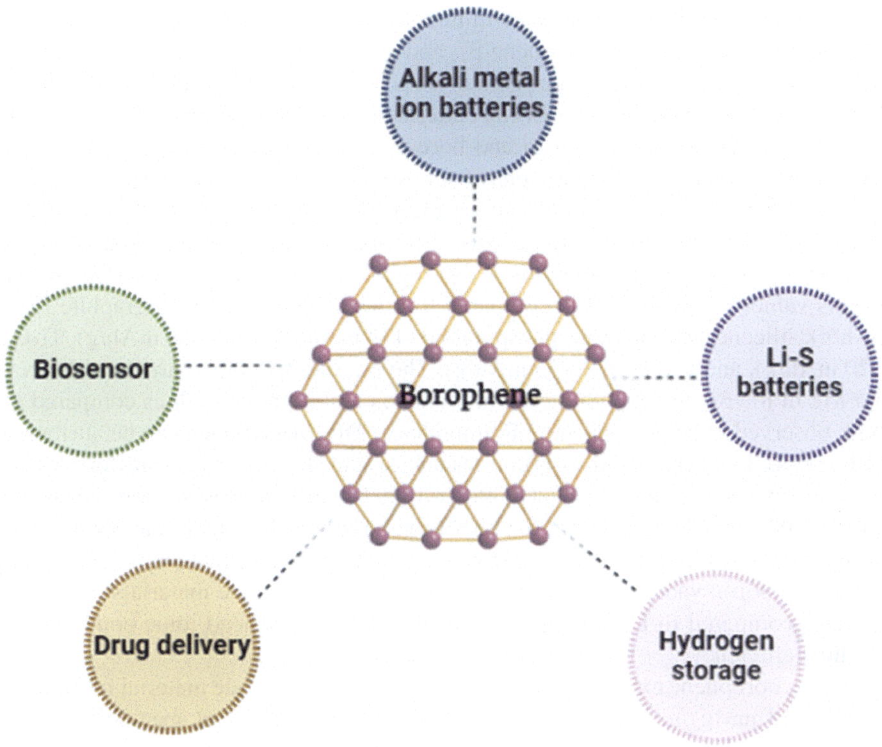

Fig. 3 Potential applications of Borophene: Alkali metal ion batteries, Li–S batteries, Hydrogen storage, drug delivery, biosensors, etc.

4.1 Alkali Metal Ion Batteries, Li–S Batteries, Hydrogen Storage

Alkali Metal Ion Batteries

Borophene demonstrates considerable potential as an anode material for lithium (Li), sodium (Na), and magnesium (Mg) ion batteries due to its notable attributes, including a high theoretical specific capacity, favorable electrical conductivity, and exceptional ion transport capabilities [22]. The exceptional ionic conductivity and outstanding electronic conductivity of borophene contribute to its ability to facilitate efficient charging and discharging processes. The 2-Pmmn phase of borophene has been employed in ion batteries containing Na, K, Mg, Ca, and Al. The 2-Pmmn phase of borophene has been seen to exhibit a very high capacity and rapid ion diffusion. The anode capacity of the borophene 2-Pmmn phase for Mg and aluminum (Al) ion batteries is reported to be 6611 mAh/g and 9917 mAh/g, respectively. The energy barriers associated with the transportation of Mg and Al ions are merely 11.76 meV and 39.24 meV, correspondingly. In addition, the β_{12} and χ_3 phases of borophene demonstrate exceptional capacity in lithium, sodium, and magnesium ion batteries. The anode capacities of the borophene-β_{12} and χ_3 phases for lithium-ion batteries are 1984 mAh/g and 1240 mAh/g, respectively. The suitability of borophene 2-Pmmn as a negative electrode material for Li-ion batteries was initially proposed by Zhao et al. The strong interaction between Li and borophene is attributed to the high surface activity exhibited by the 2-Pmmn phase of borophene. The initial interlayer voltage is measured to be 1.12 eV when the Li adsorption range is low, and it gradually drops as the Li adsorption range grows. The fully bonded phase of borophene is denoted as $Li_{0.75}B$, which exhibits a theoretical specific capacity of 1860 mAh/g. This value is notably higher compared to other materials such as graphite (372 mAh/g), silicene (954 mAh/g), phosphorene (433 mAh/g), VS_2 (466 mAh/g), Ti_3C_2 (320 mAh/g), and $Li_4Ti_5O_{12}$ (175 mAh/g). The migratory energy barriers of Li, Na, and Mg in the β_{12} and $\chi3$ phases of borophene exhibit greater values compared to those observed in the borophene 2-Pmmn phase. This discrepancy can be attributed to the periodic arrangement of boron vacancies and the structural transformations from a wavy configuration to a flat configuration. Nevertheless, when considering its application as an anode material for Li/Na-ion batteries, the energy barrier remains within acceptable limits. Hence, it can be observed that the utilization of borophene-containing boron vacancies exhibits superior efficacy as anode materials in Na-ion batteries compared to Li-ion batteries, mostly due to enhanced ionic conductivity and flow efficiency.

Hence, borophene exhibits considerable potential as an anode material for lithium Li, Na, and mMg) ion batteries owing to its notable theoretical specific capacity, superior electronic conductivity, and extraordinary ion transport characteristics. In a recent study, the resonant ball milling method and one-step molding strategy are employed to generate a stable heterostructure of borophene/graphene (B/G). The purpose of the synthesized borophene nanocomposite is to facilitate the utilization of borophene in lithium-ion batteries as shown in Fig. 4 [4]

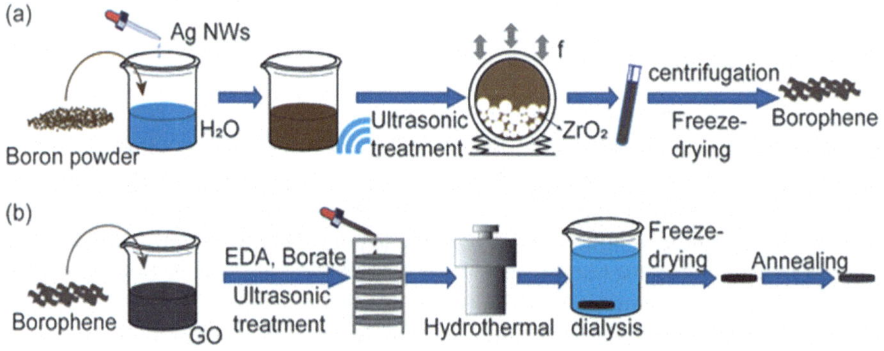

Fig. 4 Schematic representation of the preparation methodology of borophene and its. **a** borophene and **b** borophene-RGO. Reprinted (adapted) with permission from [4]. Copyright 2022 John Wiley & Sons

Li–S Batteries

Li–S batteries have garnered significant attention in recent years because of their cost-effectiveness and ample availability of sulfur, along with their impressive energy density of 2500 W h kg^{-1}. Nevertheless, there exist numerous significant obstacles that must be addressed before the practical implementation of Li–S batteries [23]. These issues include the relatively low conductivity of Li polysulfide and the substantial volume change of S during cycling, resulting in rapid power degradation and inadequate throughput performance. The primary concern of Li–S batteries is the phenomenon commonly referred to as the "shuttle effect" [23]. This phenomenon arises when the surface of the lithium metal anode becomes passivated as a result of the migration of highly soluble Li polysulfides from the cathode to the Li metal anode, resulting in undesired losses. The utilization of active materials may lead to a decrease in cycling performance [24]. Hence, the development of appropriate anchoring materials for the S cathode is imperative to effectively sequester dissolved lithium polysulfides, consequently mitigating the migration phenomenon in Li–S batteries [25]. The present study employed first-principles simulations to examine the possible utilization of borophene as an anchoring material for lithium-sulfur batteries. According to Zhao et al., the 2-Pmmn borophene phase does not exhibit strong adhesion properties when used as a protective material for the interaction between lithium polysulfide and the 2-Pmmn phase of borophene. The adsorption energies of Li_2S_4, Li_2S_6, and Li_2S_8 on the 2-Pmmn stage of borophene are reported to be 6.45, 4.32, and 6.18 eV, respectively. The significant height of the intuitive barrier might lead to the degradation of lithium polysulfide, resulting in the irreversible loss of sulfur throughout the charging and discharging processes. The adsorption energies of Li_2S_4, Li_2S_6, and Li_2S_8 on borophene in the χ_3 phase were measured to be 2.67, 2.53, and 2.87 eV, respectively. These results indicate that borophene in the χ_3 phase exhibits significant potential as an anchoring material for lithium-sulfur batteries. The presence of appropriate adsorption forces plays a crucial role in mitigating the shuttle effect and maintaining the integrity of the periodic structure during the processes of charging and discharging. Furthermore, it is worth noting that borophene demonstrates a metallic electrical structure during the entire functioning of a battery [26]. Hence, the borophene χ_3 phase exhibits considerable potential as a viable substrate material for lithium-sulfur batteries. Furthermore, the β_{12} phase of borophene demonstrates favorable performance as a suitable substrate material for lithium-sulfur batteries. The adsorption energies of Li_2S, Li_2S_2, Li_2S_4, Li_2S_6, and Li_2S_8 on the β_{12} borophene phase are reported as 3.34, 2.89, 1.45, 1.53, and 1.36 eV, respectively. Furthermore, it has been observed that the β_{12} phase of borophene demonstrates a metallic electrical structure throughout both the charging and discharging processes. Consequently, the utilization of borophene in their χ_3 and β_{12} phases has considerable potential as anchoring materials for lithium-sulfur batteries.

Hydrogen Storage

Borophene, which is recognized as the lightest 2D material, exhibits considerable promise as a material for hydrogen storage due to its exceptional capacity for storing hydrogen [27]. Numerous inquiries have been conducted to explore this subject matter. It is anticipated that forthcoming fuel cells utilizing hydrogen will provide environmentally friendly energy while exhibiting an energy efficiency that is 2.5–2.7 times greater than that of conventional fossil fuels. Nevertheless, the storage and transportation of hydrogen pose significant challenges. In the context of hydrogen storage, it is desirable to have a storage method that is both uncomplicated and cost-effective, while also exhibiting high volumetric and gravimetric density. Additionally, long-term cyclic stability, or reversibility, is a key characteristic that is sought after. 2D materials emerge as a highly promising candidate for hydrogen storage due to their ability to fulfill all the necessary criteria. To do this, borophene must have superior performance in comparison to graphene. Er et al. utilized first-principles DFT calculations employing the Vienna Ab initio Simulation Package (VASP) to make predictions regarding the hydrogen storage capacity of the planar borophene sheet. Their findings indicate an anticipated increase in the hydrogen storage capacity, estimated to be 0.15 eV/H_2. According to DFT investigations, the hydrogen storage capacity of Ca- and Li-doped borophene structures is significantly higher (0.142 and 0.176 eV/H_2, respectively) when compared to flat borophene sheets. The storage capacities of the borophene layer were enhanced by Lee et al. and Wang et al. through the incorporation of Ca atoms. This increased storage capacities ranging from 0.20 to 0.32 eV/H_2, as determined using the original VASP approach. The presence of protrusions, ridges, edges, and flaws within the borophene layer can function as potential anchoring sites for hydrogen atoms. Moreover, the combination of boron with hydrogen leads to the formation of borane gas (BH_3), rendering the borophene layer, which possesses a greater surface area compared to bulk boron, more prone to interactions with hydrogen gas.

4.2 *Biosensors*

Borophene exhibits significant potential for utilization in gas detection biosensing applications owing to its expansive surface area and notable gas molecule absorption capabilities. The monitoring capabilities of this system extend beyond non-toxic gases like ethanol (CH_2OH) and ammonia (NH_3) to encompass highly hazardous gases such as formaldehyde (CHOH), carbon monoxide (CO), and nitrogen monoxide (NO), in addition to graphene and phosphorous gas biosensors [28]. In recent times, there has been a significant surge in the interest surrounding sensors with sensitivities comparable to those of biosensors. The borophene surface exhibits a notable affinity for highly poisonous gases, owing to its strong adsorption capabilities. Furthermore, unlike other 2D materials like graphene, borophene possesses an electronic band gap, rendering it a viable candidate for biosensors targeting very toxic gases [29]. Borophene exhibits a significant drop in its electronic bandgap when gas is adsorbed, which facilitates the transfer of a substantial number of electrons and enhances its electrical conductivity. This property makes borophene highly valuable for applications involving high-risk gas biosensors. HCOH is a highly dangerous and irritating carcinogen, and contact can cause watery eyes, difficulty breathing, and asthma attacks. Nevertheless, owing to its favorable chemical reactivity and exceptional thermal stability, it finds extensive utilization across many industrial and medicinal domains. Thus, to effectively mitigate the adverse consequences of HCOH, it becomes imperative to devise HCOH biosensors with heightened sensitivity. In their study, Ansari et al. examined the viability of B_{36} as a biosensor for HCOH by conducting DFT simulations to analyze its hexagonal hollow structure. The researchers found that the presence of HCOH noticeably increases the conductivity of B_{36}. It has been observed that this phenomenon has the potential to generate electrical signals. Moreover, it can be observed that the magnitude of the electric signal amplifies proportionally with the adsorption of additional HCOH molecules. This phenomenon suggests that B36 exhibits a high level of sensitivity towards HCOH concentration (or HCOH pressure), hence establishing its potential as a biosensor for HCOH. Ta. In this context, B_{36} borophene reliably plays the role of a semiconductor with high conductivity and generates electrical signals in the presence of HCOH. Huang et al. investigated the transport characteristics of borophene after the adsorption of CO or NO gases, utilizing a NEGF-based methodology. Based on the findings, it is evident that the 2-Pmmn phase exhibited notably superior adsorption capabilities for CO, NO, NO_2, and NH_3 in comparison to graphene, MoS_2, and phosphorus. A notable alteration in the transfer function is observed with the adsorption of CO, NO, NO_2, or NH_3, in contrast to the initial 2-Pmmn phase of borophene. Furthermore, there is an observable reduction in the intensity of the current–voltage (I–V) pattern as a result of the adsorption of gas. The aforementioned qualities exemplify the alternating states of the sensor system. In the end, the application of borophene to enhance 2D nanomaterials devices was recently reported, and the hexagonal structure of B_{36} borophene exhibits excellent properties, making it ideal for integrating 2D nanostructures into biometric sensors.

4.3 Drug Delivery

The trick to minimizing the negative effects of chemotherapy is to release the chemotherapy drugs locally. It is important to develop intelligent drug delivery platforms that are responsive to the tumor microenvironment. Borophene exhibits a significantly expansive specific surface area, hence facilitating the accommodation of drugs and enabling the attachment of functional groups. This characteristic renders borophene highly appropriate for a wide range of therapeutic applications in the field of anticancer treatment. To establish a continuous distribution of chemotherapeutic medications within the body, Ji et al. conducted an experimental study whereby doxorubicin (DOX) was subjected to incubation with borophene in a drug delivery system based on borophene. The loading rate for this system was determined to be 114%. We then coated the system with PEG-NH_2 and demonstrated successful PEG modification and DOX loading. Borophene, in its capacity as a drug carrier, had favorable pH characteristics and notable photothermal reactivity. The DOX release rate by borophene was 8.3% and 24% at pH 7.4 and pH 5.0, respectively. Due to the much lower pH levels observed in the cancer microenvironment compared to normal tissues, the utilization of borophene has been found to facilitate precise and targeted medication delivery specifically to tumor sites. Moreover, borophene exhibits potential as a carrier for near-infrared (NIR) reactions in the field of medication delivery. In the aforementioned experimental conditions, the application of a 5-min NIR irradiation increased the drug release rate. Specifically, at pH 7.4, the rate at which the drug was released reached 41.4%, while at pH 5.0, it jumped to 77.6%. In brief, borophene exhibits notable attributes such as substantial loading of drug capacity, pH responsiveness, and dual infrared reactivity. These characteristics make it a promising candidate for the development of an integrated drug delivery system that can effectively cater to both cancer detection and cancer therapy [30].

4.4 Borophene as Supercapacitor

Recent theoretical investigations have demonstrated that borophene when utilized as a 2D material, has superior electrochemical characteristics compared to graphene. Consequently, borophene emerges as a more auspicious candidate for implementation as an electrode material in batteries and supercapacitors. Several layers of B-sheets prepared by ultrasound-assisted liquid-phase exfoliation were used as electrode materials in supercapacitors. The exfoliated multilayer boron plate exhibits remarkable stability against air oxidation because it is protected from residual solvents [31]. The recorded data indicates a broad potential range of up to 3.0 V and a substantial energy density of 46.1 Wh/Kg. The cycling stability of the original specific capacity is deemed excellent, as evidenced by a capacity retention rate of around 88.7% after 6000 cycles. The present experimental study demonstrates the controlled synthesis of a limited quantity of stacked boron layers, hence broadening the scope for prospective practical implementations of boron layers.

4.5 Borophene Within Nanotechnologies

Borophene possesses excellent electrical conductivity and the ability to adjust its bandgap, rendering it highly suitable for advanced nanoelectronics applications. Researchers have demonstrated its potential for use in next-generation transistors and sensors [3]. Its attractiveness in nanoelectronics applications is enhanced by the ability to control its electronic properties through voltage and doping. Its mechanical flexibility and stiffness have attracted the interest of researchers for its application in nanomechanical sensors that can sense a variety of physical and chemical inputs. Its unique mechanical properties make it a promising option for highly sensitive and multifunctional sensors [32]. These applications demonstrate the applicability and potential of borophene in nanotechnology and provide prospects for innovation in fields such as electronics, sensing, etc. [33].

4.6 Borophene as a Catalyst

The researchers have shown that borophene exhibits remarkable catalytic prowess in hydrogen and oxygen production reactions, as well as in the electroreduction reaction of CO_2. This discovery holds promising implications for the development of novel energy cycles centered around hydrogen and water. Borophene exhibits many potential uses in the field of hydrogen evolution (HER) [34]. Borophene has been identified as the catalyst with the lowest weight for the HER. As per the findings of Wang et al., borophene exhibits exceptional performance as a catalyst for the HER, characterized by an extremely low hydrogen adsorption-free energy, metallic conductivity, and a substantial quantity of active sites inside its basal plane [35]. The hydrogen atom has a significant interaction with borophene 2-Pmmn due to its high surface activity. In contrast, the free energy associated with the adsorption of hydrogen (GH) on borophene is merely 0.02 eV, which is significantly lower compared to the free energy of GH hydrogen adsorption on Pt as a catalyst, which measures 0.09 eV. Small hydrogen adsorption free energy GH was also found in leaves and phase 1 of borophene. In the borophene sheet and phase 1 of borophene, a small GH adsorption-free energy of hydrogen was also determined [36]. In addition, the silver substrate did not affect the strong catalytic activity of borophene. Borophene, when negatively charged, can absorb a surface area of approximately 6.731014 cm^2 of CO_2. In the negatively charged phase of borophene, reversible CO_2 capture/release reactions are observed. Control of charging and discharging involves the activation and deactivation of charges in the borophene phase. Sun et al. also presented a feasible technique to fabricate new CO_2 reduction electrocatalysts. The researchers demonstrated that the inclusion of secondary adsorption sites enhances the catalytic performance of borophene-supported Cu chains in the electro-reduction of CO_2. This results in a reduction in the minimum potential required for the CO_2-CH_3OH reaction pathway, indicating a strong catalytic activity [37, 38].

Borophene-based materials have a wide range of applications, including those in the fields of energy, sensors, supercapacitors, biomedicine, electrode reduction, and batteries. The applications are graphically represented in Fig. 5 [39]. The advanced sectors utilizing the potential of borophene are given in Table 3.

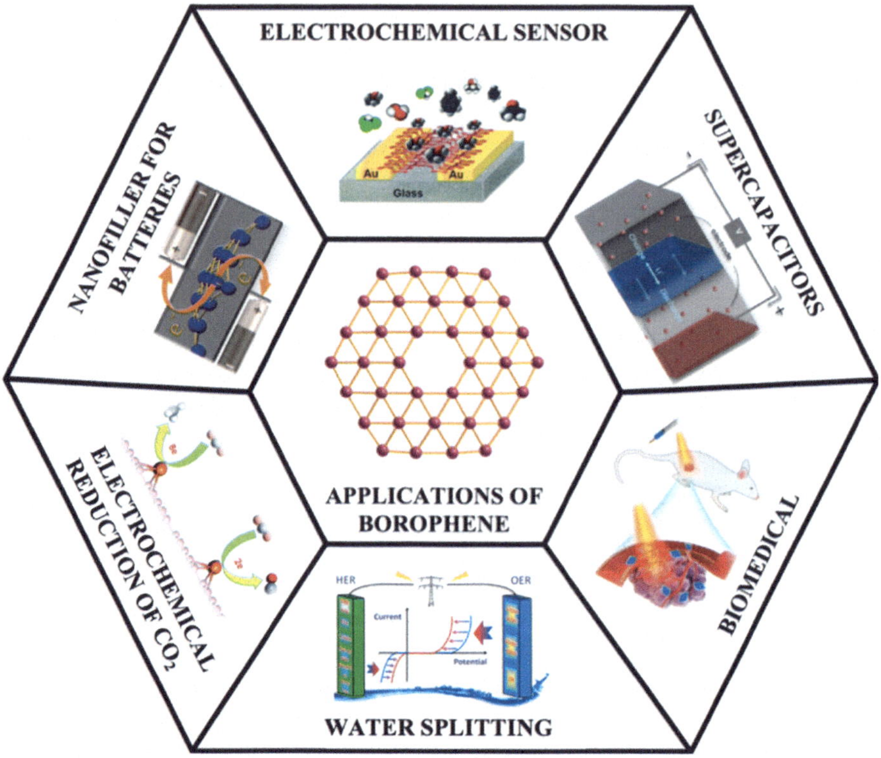

Fig. 5 Borophene-based materials have diverse uses in energy, sensors, supercapacitors, biomedical devices, electrode reduction, and batteries. Reprinted (adapted) with permission from [39]. Copyright 2024 Elsevier

5 Conclusion

2D-borophene is a versatile nanomaterial with superior optical, electrical, mechanical, thermal, and magnetic capabilities. It can replace traditional materials in biomedical sensors, diagnostic instruments, and high-performance healthcare equipment. This chapter examines the computational and experimental aspects of chemistry that are necessary to improve the characteristics of borophene. It highlights its high electrical conductivity, chemistry-oriented biocompatibility, and photo-induced features, making it suitable for biosensing, bioimaging, cancer treatment, and theragnostic. The chapter concludes with advancements in borophene-based energy storage, hydrogen storage, supercapacitors, gas sensors, catalysis, electronics, and biomedical applications.

Table 3 Comparison of various applications of borophene

Application	Descriptions	Potential applications
Energy storage	Storing and releasing energy efficiently	Negative electrode material for high-performance batteries. High-performance hydrogen storage material The potential of next-generation energy storage
Therapeutics	Medical and pharmaceutical applications	Drug delivery systems due to the biocompatibility of borophene Potential for targeted drug delivery and Bio-sensing, and diagnostics application
Nanotechnologies	Advanced technologies at the nanoscale	It can be used in transistors, interconnects, and other electronic components, potentially leading to faster and smaller devices
Supercapacitors	High-capacity capacitors for energy storage	High surface area and conductivity make borophene suitable for supercapacitors. Potential for high-energy–density supercapacitors
Catalyst	A substance that speeds up chemical reactions	Potential as a catalyst in various chemical reactions. Electrochemical catalysis for fuel cells and energy conversion. Improved catalytic activity due to unique electronic structure

Acknowledgements Mohd. Abubakar Sadique, Shalu Yadav, and Raju Khan express their gratitude to the Director of CSIR-AMPRI Bhopal, India, for his interest in this work. Shalu Yadav expresses gratitude to the CSIR in India for the SRF award.

References

1. Mannix, A.J., Zhou, X.F., Kiraly, B., Wood, J.D., Alducin, D., Myers, B.D., Guisinger, N.P.: Synthesis of borophenes: anisotropic, two-dimensional boron polymorphs. Science **350**(6267), 1513–1516 (2015)
2. Ibrahim, M.A., Mahmoud, A.H., Soliman, K.A., Mekhemer, G.A., Ahmed, M.N., Shawky, A.M., Moussa, N.A.: Borophene and pristine graphene 2D sheets as potential surfaces for the adsorption of electron-rich and electron-deficient π-systems: a comparative DFT study. Nanomaterials **12**(6), 1028 (2022)
3. Mannix, A.J., Zhang, Z., Guisinger, N.P., Yakobson, B.I., Hersam, M.C.: Borophene as a prototype for synthetic 2D materials development. Nat. Nanotechnol. **13**(6), 444–450 (2018)
4. Yu, J., Zhou, M., Yang, M., Zhang, Y., Xu, B., Li, X., Tao, H.: Pristine and defective 2D borophene/graphene heterostructure as the potential anode of lithium-ion batteries. Adv. Mater. Interfaces **9**(12), 2102088 (2022)
5. Feng, B., Zhang, J., Zhong, Q., Li, W., Li, S., Li, H., Wu, K.: Experimental realization of two-dimensional boron sheets. Nat. Chem. **8**(6), 563–568 (2016)
6. Zhou, Y.P., Jiang, J.W.: Molecular dynamics simulations for mechanical properties of borophene: parameterization of valence force field model and Stillinger-Weber potential. Sci. Rep. **7**(1), 45516 (2017)
7. Zhang, Z., Yang, Y., Penev, E.S., Yakobson, B.I.: Elasticity, flexibility, and ideal strength of borophenes. Adv. Func. Mater. **27**(9), 1605059 (2017)
8. Ranjan, P., Sahu, T.K., Bhushan, R., Yamijala, S.S., Late, D.J., Kumar, P., Vinu, A.: Freestanding borophene and its hybrids. Adv. Mater. **31**(27), 1900353 (2019)

9. Mortazavi, B., Rahaman, O., Dianat, A., Rabczuk, T.: Mechanical responses of borophene sheets: a first-principles study. Phys. Chem. Chem. Phys. **18**(39), 27405–27413 (2016)
10. Li, D., Gao, J., Cheng, P., He, J., Yin, Y., Hu, Y., Chen, L., Cheng, Y., Zhao, J.: 2D boron sheets: structure, growth, and electronic and thermal transport properties. Adv. Func. Mater. **30**(8), 1904349 (2020)
11. Xiao, H., Cao, W., Ouyang, T., Guo, S., He, C., Zhong, J.: Lattice thermal conductivity of borophene from first principle calculation. Sci. Rep. **7**(1), 45986 (2017)
12. Bhavyashree, M., Rondiya, S.R., Hareesh, K.: Exploring the emerging applications of the advanced 2-dimensional material borophene with its unique properties. RSC Adv. **12**(19), 12166–12192 (2022)
13. Lherbier, A., Botello-Méndez, A.R., Charlier, J.-C.: Electronic and optical properties of pristine and oxidized borophene. 2D Materials **3**(4), 045006 (2016)
14. Wang, Z., Lü, T.Y., Wang, H.Q., Feng, Y.P., Zheng, J.C.: New crystal structure prediction of fully hydrogenated borophene by first principles calculations. Scientific Reports **7**(1), 609 (2017)
15. Er, S., de Wijs, G.A., Brocks, G.: DFT study of planar boron sheets: a new template for hydrogen storage. J. Phys. Chem. C **113**(43), 18962–18967 (2009)
16. Adamska, L., Sharifzadeh, S.: Fine-tuning the optoelectronic properties of freestanding borophene by strain. ACS Omega **2**(11), 8290–8299 (2017)
17. Adamska, L., Sadasivam, S., Foley IV, J.J., Darancet, P., Sharifzadeh, S.: First-principles investigation of borophene as a monolayer transparent conductor. J. Phys. Chem. C **122**, acs.jpcc.7b10197, Iv (2018)
18. Adamska, L., Sadasivam, S., Foley, J.J., Darancet, P., Sharifzadeh, S.: First-principles investigation of borophene as a monolayer transparent conductor. J. Phys. Chem. C **122**(7), 4037–4045 (2018)
19. Penev, E.S., Kutana, A., Yakobson, B.I., Boron, C.T.-D.: Can two-dimensional boron superconduct? Nano Lett. **16**(4), 2522–2526 (2016)
20. Li, G., Zhao, Y., Zeng, S., Zulfiqar, M., Ni, J.: Strain effect on the superconductivity in borophenes. J. Phys. Chem. C **122**(29), 16916–16924 (2018)
21. Zhao, Y., Zeng, S., Ni, J.: Phonon-mediated superconductivity in borophenes. Appl. Phys. Lett. **108**(24), 242601 (2016)
22. Jiang, H.R., Lu, Z., Wu, M.C., Ciucci, F., Zhao, T.S.: Borophene: A promising anode material offering high specific capacity and high-rate capability for lithium-ion batteries. Nano Energy **23**, 97–104 (2016)
23. Zhang, L., Liang, P., Shu, H.B., Man, X.L., Li, F., Huang, J., Dong, Q.M., Chao, D.L.: Borophene as efficient sulfur hosts for lithium–sulfur batteries: Suppressing shuttle effect and improving conductivity. J. Phys. Chem. C **121**(29), 15549–15555 (2017)
24. Seh, Z.W., Sun, Y., Zhang, Q., Cui, Y.: Designing high-energy lithium–sulfur batteries. Chem. Soc. Rev. **45**(20), 5605–5634 (2016)
25. Jiang, H.R., Shyy, W., Liu, M., Ren, Y.X., Zhao, T.S.: Borophene and defective borophene as potential anchoring materials for lithium–sulfur batteries: a first principles study. J. Mater. Chem. A **6**(5), 2107–2114 (2018)
26. Li, Q.F., Duan, C.G., Wan, X.G., Kuo, J.L.: Theoretical prediction of anode materials in Li-ion batteries on layered black and blue phosphorus. J. Phys. Chem. C **119**(16), 8662–8670 (2015)
27. Wang, J., Du, Y., Sun, L.: Ca-decorated novel boron sheet: a potential hydrogen storage medium. Int. J. Hydrogen Energy **41**(10), 5276–5283 (2016)
28. Huang, C.-S., Murat, A., Babar, V., Montes, E., Schwingenschlögl, U.: Adsorption of the gas molecules NH3, NO, NO2, and CO on borophene. J. Phys. Chem. C **122**(26), 14665–14670 (2018)
29. Shukla, V., Wärnå, J., Jena, N.K., Grigoriev, A., Ahuja, R.: Toward the realization of 2D borophene based gas sensor. J. Phys. Chem. C **121**(48), 26869–26876 (2017)
30. Ji, X., Kong, N., Wang, J., Li, W., Xiao, Y., Gan, S.T., Zhang, Y., Li, Y., Song, X., Xiong, Q., Shi, S., Li, Z., Tao, W., Zhang, H., Mei, L., Shi, J.: A novel Top-Down synthesis of ultrathin 2D boron nanosheets for multimodal Imaging-Guided cancer therapy. Adv. Mater. **30**(36), 1803031 (2018)

31. Li, H.L., Jing, L., Liu, W.W., Lin, J.J., Tay, R.Y., Tsang, S.H., Teo, E.H.T.: Scalable production of few layer boron sheets by liquid-phase exfoliation and their superior super capacitive performance. ACS Nano **12**(2), 1262–1272 (2018)
32. Shi, L., Ling, C., Ouyang, Y., Wang, J.: High intrinsic catalytic activity of two-dimensional boron monolayers for the hydrogen evolution reaction. Nanoscale **9**(2), 533–537 (2017)
33. Pei, A., Zheng, G., Shi, F., Li, Y., Cui, Y.: Nanoscale nucleation and growth of electrodeposited lithium metal. Nano Lett. **17**(2), 1132–1139 (2017)
34. Mir, S.H., Chakraborty, S., Jha, P.C., Wärnå, J., Soni, H., Jha, P.K., Ahuja, R.: Two-dimensional boron: lightest catalyst for hydrogen and oxygen evolution reaction. Appl. Phys. Lett. **109**(5), 053903 (2016)
35. Wang, Z.Q., Lü, T.-Y., Wang, H.-Q., Feng, Y. P., Zheng, J.-C.: Review of borophene and its potential applications. Front. Phys. **14**(3), 33403 (2019)
36. Singh, Y., Back, S., Jung, Y.: Computational exploration of borophane-supported single transition metal atoms as potential oxygen reduction and evolution electrocatalysts. Phys. Chem. Chem. Phys. **20**(32), 21095–21104 (2018)
37. Nørskov, J.K., Bligaard, T., Logadottir, A., Kitchin, J.R., Chen, J.G., Pandelov, S., Stimming, U.: Trends in the exchange current for hydrogen evolution. J. Electrochem. Soc. **152**(3), J23 (2005)
38. Liu, C.W., Dai, Z.X., Zhang, J., Jin, Y.G., Li, D.S., Sun, C.H.: Two-dimensional as metal free catalysts for hydrogen evolution reaction. J. Phys. Chem. C **122**(33), 19051–19055 (2018)
39. Nakai, A., Rajeev, R., Varghese, A.: Synthesis, properties, and state-of-the-art advances in surface tuning of borophene for emerging applications. Mater. Today Sustain. 100743 (2024)

Synthesis Strategies of 2D Boron Nanosheets: Computational or Theoretical Route and Experimental Approaches

Monu Kumar Shukla, Harshita Tiwari, and Raju Khan

Abstract Two-dimensional (2D) materials have garnered significant interest in recent years due to their unique and outstanding electrical, mechanical, and optical properties. 2D boron nanosheets have emerged as promising candidates for a variety of applications due to their unique and tunable properties. This book chapter provides a comprehensive overview of the methods used to create and develop 2D boron nanosheets. The chapter discusses the theoretical and computational methods that have been used to study 2D boron nanosheets. We examine the theoretical underpinnings of the prediction of the boron structure, emphasizing the significance of first-principles computations and molecular dynamics simulations. The topic effortlessly shifts to experimental procedures, including chemical vapor deposition, physical vapor deposition, and solution-based methods used to construct 2D boron nanosheets. Additionally, we highlighted the connections between synthesis, structure, and property and provided insights into the difficulties and opportunities of designing 2D boron materials for a variety of applications. This chapter tries to bridge the gap between theoretical predictions and experimental reality by offering a thorough overview of the most recent techniques for creating 2D boron nanosheets.

Keywords 2D material synthesis · Boron nanomaterials · Crystal growth · Chemical vapor deposition · Thin film deposition

M. K. Shukla (✉) · H. Tiwari
Drug Chemistry Research Centre, Kanadia Road, Indore, Madhya Pradesh 452003, India
e-mail: drugchemistryresearchcenter@gmail.com

R. Khan
Industrial Waste Utilization, Nano and Biomaterials, CSIR-Advanced Materials and Processes Research Institute (AMPRI), Hoshangabad Road, Bhopal, Madhya Pradesh 462026, India

Academy of Scientific and Innovative Research (AcSIR), Ghaziabad 201002, India

R. Khan et al. (eds.), *2D Boron Nanosheets*, Engineering Materials,
https://doi.org/10.1007/978-981-97-7014-4_4

1 Introduction

Two-dimensional (2D) nanosheets are materials that exist in a two-dimensional plane, typically with only a few atomic or molecular layers in thickness. These materials are often composed of atoms or molecules arranged in a single layer, giving them unique properties and a wide range of potential applications in various fields such as including electronics, optics, and nanotechnology. Some of the 2D nanosheets include graphene, transition metal dichalcogenides (TMDs), black phosphorus (phosphorene), hexagonal boron nitride (h-BN): hexagonal boron nitride and silicene, etc.

2D boron nanosheets are an interesting and rapidly evolving family of nanomaterials. These materials exhibited exceptional electrical, mechanical, and chemical properties due to their flat, single-layer boron atom structure [1, 2]. The discovery and study of 2D boron nanosheets have expanded applications in electronics, optoelectronics, energy storage, catalysis, and various biomedical fields. Like graphene, 2D boron nanosheets create single-layer sheets. In its hexagonal lattice, each boron atom is bonded to three neighbors. A planar structure with intriguing electrical properties results from this design [3].

Lin and Connell, 2012 reported a controlled version of 2D boron nanosheets which have been synthesized using a variety of methods, including chemical vapor deposition (CVD), mechanical exfoliation, electrochemical synthesis, and laser ablation. They also mentioned that the density functional theory (DFT) and molecular dynamics (MD) simulations have been indispensable for predicting the stable structures of 2D boron nanosheets and guiding the experimental efforts to synthesize them [4].

Though 2D boron nanosheets lack a full valence electron shell which develops stability issues. 2D boron nanosheets may have different bonding arrangements and structural changes. Researchers are testing these stabilized materials in various circumstances [5]. Also, 2D boron nanosheets exhibit unique electrical properties that distinguish them from other 2D materials. Depending on structure and doping, boron nanosheets may act metallically or semiconducting due to their peculiar atom arrangement. These properties make them promising for future electrical and optoelectronic devices [6, 7]. 2D boron nanosheets are ideal for composite reinforcement due to their mechanical strength and flexibility. Their chemical reactivity and catalytic activity for chemical processes and energy-related activities have also gained attention [8].

Zhao et al., 2019 investigated the chemical composition and structure of these boron nanosheets (Fig. 1), with the help of a series of rigorous analytical techniques and spectroscopic methods employed, including X-ray photoelectron spectroscopy (XPS), transmission electron microscopy (TEM), Raman spectroscopy, and Fourier-transform infrared spectroscopy (FTIR). These complementary approaches provided crucial insights into the elemental composition, atomic arrangement, and bonding characteristics, allowing us to gain a comprehensive understanding of the unique properties and potential applications of these nanosheets in various fields, from materials science to nanoelectronics [9].

2D boron nanosheets are a novel material class with a wide range of potential applications. Their unique properties, such as their high electrical conductivity, high thermal conductivity, and chemical inertness, make them promising candidates for

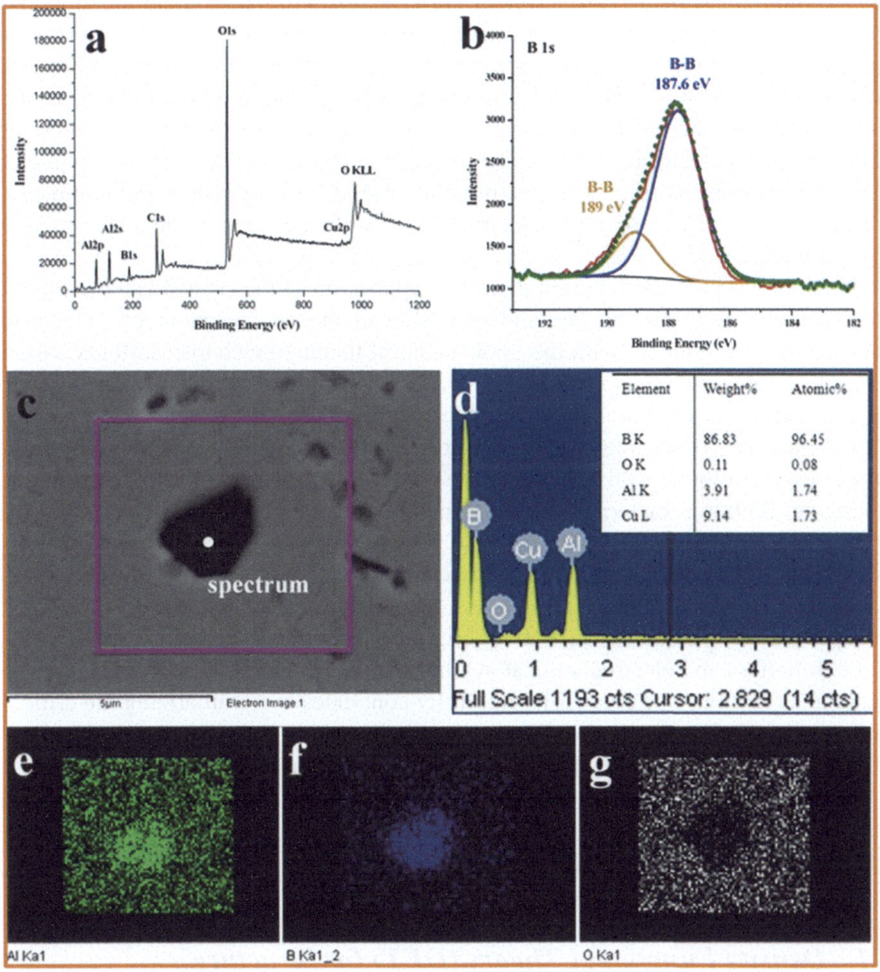

Fig. 1 Boron nanosheet composition. **a**, **b** Boron nanosheet XPS findings. Al, O, and Cu are dissolved in boron sheets, but **b** shows pure B-B peaks with olive circles after fitting blue and orange curves; **c**–**g** EDS findings of the boron phase in heated Cu-B powders without grinding. With almost 96% boron, this technique provides suggestive evidence of nanosheet composition. According to **e**–**g**, element Al dissolves in boron without oxidation. Reproduced from Ref. [9] with permission under the Creative Commons CC BY license from Springer Nature, copyright 2019

use in a variety of applications. They might be catalysts for chemical reactions, transistors in electronics, and energy storage devices like batteries and supercapacitors. Their unique electrical structure makes them valuable for spintronics and quantum computing. 2D boron nanosheets, current research focuses on innovative synthetic routes, their fundamental properties, and commercial scalability [10–13].

Two primary categories of 2D boron nanosheets are borophene and boron nitride. Borophene represents a single layer of boron atoms with remarkable metallic and electronic properties, making it a promising candidate for next-generation electronics and catalysis. On the other hand, boron nitride, consisting of alternating boron and nitrogen atoms, exhibits excellent insulating properties, high thermal conductivity, and exceptional chemical stability, positioning it as an ideal material for thermal management in electronics and various protective coatings. These distinct types of 2D boron nanosheets offer a tantalizing playground for researchers and engineers, driving innovation across a wide spectrum of scientific disciplines and technological applications [14, 15].

This chapter provides a comprehensive exploration of the critical computational and synthetic methods that are indispensable in the advancement of 2D boron nanosheets. In alignment with the book's central theme, the chapter unfolds with a focus on computational approaches in Sect. 2, where it delves into the utility of DFT for predicting the structures and properties of these nanosheets. It further explores MD simulations, shedding light on their significance in understanding dynamic processes. Additionally, this section delves into the growth mechanisms and stability studies of 2D boron materials. Subsections under this category delve into thermodynamics, kinetics, and the simulation of growth pathways, enriching our comprehension of the synthesis processes. The chapter also addresses the computational search for optimal 2D boron nanosheets. Moving forward, it extends its coverage to experimental methods in Sect. 3, highlighting techniques such as CVD, mechanical exfoliation, and electrochemical synthesis, providing practical insights into the material's creation. The chapter thoughtfully concludes by summarizing the critical insights and challenges, reinforcing its vital role within the book's broader context of computational and experimental methods in advancing 2D boron nanosheets.

2 Computational or Theoretical Route

2.1 Density Functional Theory (DFT) for Structure Prediction

DFT has emerged as a powerful tool for predicting the structural properties of 2D boron nanosheets, opening up new horizons in materials science (Fig. 2) [16]. DFT allows researchers to explore the intricate atomic arrangements of these fascinating materials with remarkable precision, unveiling their unique electronic and mechanical characteristics [17]. By simulating the interactions between boron atoms, DFT

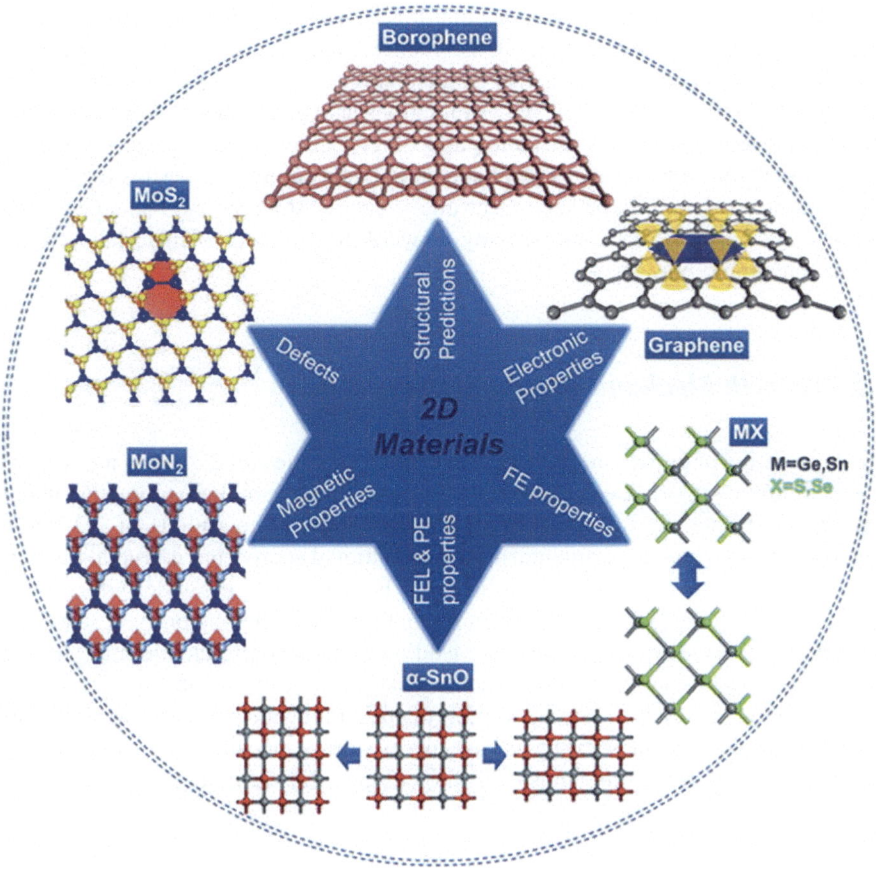

Fig. 2 Computational design of 2D boron nanosheets and property predictions for such structures. Adopted with permission from [16], copyright Elsevier 2018

aids in the discovery of stable configurations and provides valuable insights into the structural stability and electronic properties of 2D boron nanosheets. This computational approach not only accelerates the design of novel materials but also paves the way for their potential applications in electronics, catalysis, and beyond, advancing our understanding of the cutting-edge frontier of nanomaterials [18, 19].

2.2 *Applicability of Molecular Dynamics (MD) Simulations*

MD simulations have emerged as an indispensable tool in the study of 2D boron nanosheets, enabling researchers to gain deep insights into their dynamic behavior and properties at the atomic and molecular levels. Through MD simulations, scientists can unravel the intricate thermal, mechanical, and structural properties of these materials, providing a comprehensive understanding of their stability, flexibility, and response to external stimuli [20]. MD simulations also offer a platform for exploring the interactions between boron atoms and other elements, facilitating the design and optimization of novel 2D boron nanosheets with tailored properties for diverse applications, such as in advanced electronics, energy storage, and nanoelectromechanical systems (NEMS), thus driving forward the frontiers of materials science and nanotechnology [21].

2.3 *Growth Mechanisms and Stability Studies*

The exploration of growth mechanisms and stability studies for 2D boron nanosheets is of paramount importance in advancing our understanding of these promising materials. Investigating the formation processes and stability conditions of 2D boron materials through both experimental and computational approaches is crucial for their controlled synthesis and practical applications. By uncovering the underlying growth mechanisms and characterizing the factors that influence their stability, researchers can fine-tune the synthesis conditions, leading to more reproducible and tailored nanosheets. Additionally, stability studies shed light on the durability and performance of these materials in various environments, guiding their potential applications in fields such as electronics, catalysis, and energy storage, and ultimately accelerating the integration of 2D boron nanosheets into cutting-edge technologies [22, 23].

2.4 *Monte Carlo Methods*

2.4.1 Thermodynamics and Kinetics of 2D Boron Formation

Exploring the thermodynamics and kinetics of 2D boron formation is essential for deciphering the intricate processes governing the synthesis and stability of these unique nanomaterials. Thermodynamic investigations help us understand the feasibility and energetics of 2D boron formation, shedding light on the most favorable conditions for their growth [24]. Simultaneously, kinetic studies provide valuable insights into the rates at which these materials form, offering crucial information for optimizing synthesis techniques and controlling the nucleation and growth processes [25]. Together, these studies empower researchers to fine-tune the fabrication of 2D boron structures with tailored properties, enabling their integration into a wide range of applications, from advanced electronics to energy storage systems, and fostering advancements at the intersection of materials science and nanotechnology.

2.4.2 Simulating Growth Pathways

Simulating the growth pathways of 2D boron structures is a pivotal endeavor at the intersection of materials science and computational research. Through advanced computational models and simulations, scientists can unravel the intricate steps involved in the formation and evolution of these nanomaterials, shedding light on their stability, morphology, and potential applications [26]. By meticulously tracking the trajectories of boron atoms as they come together to create 2D structures, researchers gain insights into the nucleation, growth, and stabilization processes. This knowledge is invaluable for optimizing synthesis techniques, predicting the emergence of novel structural motifs, and tailoring the properties of 2D boron materials for specific technological applications, paving the way for innovation in nanotechnology and beyond [27].

2.5 Computational Search for Optimal Structures of 2D Boron Nanosheets

The computational search for optimal structures of 2D boron nanosheets represents a crucial frontier in materials science and nanotechnology. Leveraging advanced algorithms and simulation techniques, researchers explore a vast design space to identify the most stable and functional configurations of these unique materials. By calculating the energy landscapes and properties of various atomic arrangements, computational methods enable the prediction of novel 2D boron structures with tailored electronic, mechanical, and thermal properties [28]. This approach not only accelerates the discovery of promising candidates for a wide range of applications, including electronics, catalysis, and energy storage but also contributes to a deeper understanding of the fundamental principles governing the behavior of boron at the nanoscale, driving innovation and technological advancements [29].

3 Experimental Approaches for 2D Boron Nanosheets

The synthesis and characterization of 2D boron nanosheets, such as borophene, boron nanosheets, and boron nanotubes, present particular difficulties due to boron's intrinsic reactivity and structural instability in its elemental form. Nevertheless, several experimental techniques have been developed to produce and research these fascinating compounds.

3.1 Chemical Vapor Deposition (CVD)

To create the amazing class of materials known as 2D boron nanosheets, a versatile method known as CVD was used. CVD involves the introduction of boron-containing precursors at high temperatures into a controlled reaction chamber, such as boron trichloride (BCl_3) or diborane (B_2H_6) [15]. Deokar et al. described that the precursor molecules are separated from the nanosheets, and boron atoms are then deposited on the substrate surface. The development and structural characteristics of the resultant nanosheets are influenced by the substrate choice, which is crucial. The boron nanosheets are created when the boron atoms build up on the substrate and self-organize into a 2D lattice. Temperature, pressure, and precursor flow rate adjustments may all be used to precisely regulate the growth process. Scalability and control over the thickness, crystallinity, and shape of the nanosheets are two benefits of CVD. The capacity to create and investigate 2D boron nanosheets for possible use in electronics, catalysis, and materials science has greatly improved, thanks to this technique [30, 31].

3.2 Mechanical Exfoliation

Gautam and Chelliah, 2021 described an intriguing method that was motivated by the successful extraction of graphene from bulk graphite is the mechanical exfoliation production of 2D boron nanosheets. This technique involves applying mechanical pressures, such as scotch tape or mechanical cleavage, to bulk boron crystals or boron-rich materials to separate and isolate thin layers of boron. These sheets of boron, which have distinct electrical and structural characteristics, may be used to create layers that are a few atoms thick [32]. Wang et al. explained the advantages of mechanical exfoliation are its ease of use and low need for specialist equipment. Due to the brittleness of boron materials and the possibility of flaws in the resultant nanosheets, it may be difficult, however. Nevertheless, this approach has shed light on the basic characteristics of 2D boron materials and is a useful tool for investigating their prospective uses in nanoelectronics and materials research [33]. Zhao et al., 2019 prepared boron nanosheets with the help of atomization and ball milling which are the processes of mechanical exfoliation, in which bulk boron in copper is broken into pieces (Fig. 3) [9].

3.3 Electrochemical Synthesis

A potential approach that enables the controlled development of these materials on conductive substrates is the electrodeposition synthesis of 2D boron nanosheets. According to Wood et al., 2021 in this method, a boron-containing electrolyte solution

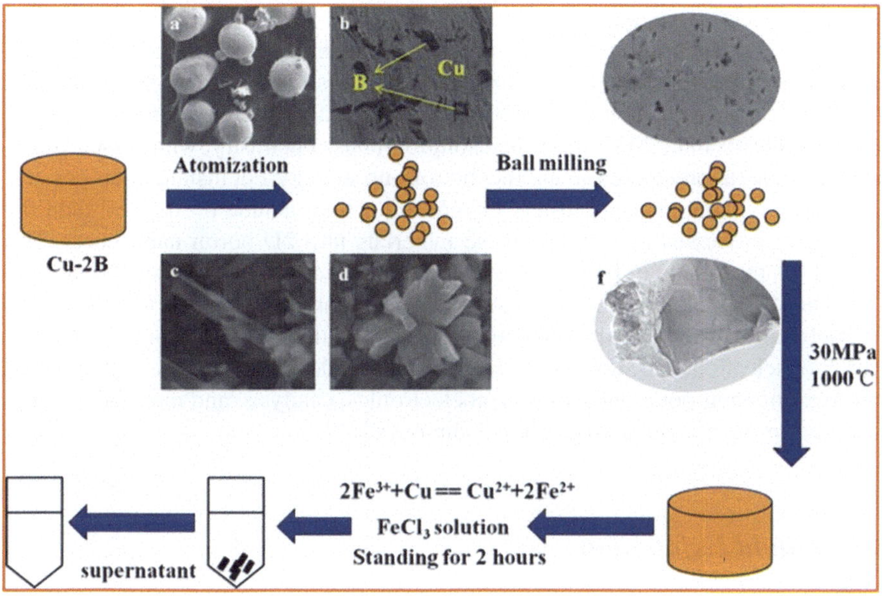

Fig. 3 Preparation of boron nanosheets. Atomization and ball milling are the processes of mechanical exfoliation, by which bulk boron in copper is broken into pieces. Reproduced from ref. 9 with permission under the Creative Commons CC BY license from Springer Nature, copyright 2019

is created, often using boron-containing chemicals, and an electrode is created using a conductive substrate, such as a metal or conductive glass. Boron ions are decreased at the electrode surface when an electric current is applied, which causes the production of boron nanosheets or other 2D boron structures [34]. Jin et al. also explained that by altering variables like electrode potential, current density, and deposition duration, electrodeposition enables exact control over the development process. The size, thickness, and shape of the resultant 2D boron nanosheets may all be altered using this technique. Furthermore, electrodeposition is a feasible and scalable method for creating 2D boron materials for diverse applications in electronics, energy storage, and catalysis. Electrodeposition may be carried out under comparatively moderate circumstances [35].

3.4 Bottom-Up Chemical Synthesis Method for 2D Boron Nanosheets

The controlled creation of these materials from boron-containing precursor molecules constitutes the bottom-up chemical synthesis approach for 2D boron nanosheets. This method enables the exact design and construction of 2D boron nanosheets with predetermined characteristics. CVD, a popular technique, involves adding a substrate to a reaction chamber along with boron-containing gases such as BCl_3 or B_2H_6. The boron-containing precursors break down at a regulated temperature and pressure, releasing boron atoms onto the substrate where they self-organize into a 2D lattice to create boron nanosheets or other 2D structures [36, 37].

Chakrabarty et al., 2022, described solution-phase chemistry, which uses boron-rich molecules as precursors, as another bottom-up strategy. For instance, it is possible to chemically create boron clusters or molecules that include boron, and then use subsequent processes to exfoliate these materials into 2D boron nanosheets. This method has the benefit of being scalable and flexible for different substrate materials [38]. The size, thickness, and structure of the resultant 2D boron nanosheets may be precisely tailored in both situations by carefully regulating the reaction conditions, precursor selection, and substrate characteristics. Understanding these materials and investigating their potential uses in nanoelectronics, catalysis, and materials science need bottom-up chemical production techniques [39].

3.5 *Liquid Exfoliation*

Liquid exfoliation has emerged as a promising method for the production of 2D boron nanosheets. This technique involves the dispersion of bulk boron materials into a solvent, followed by mechanical or ultrasonic exfoliation to create thin boron nanosheets [40]. The process results in the formation of boron layers with a high aspect ratio. Liquid exfoliation is known for its scalability and simplicity, making it a practical choice for large-scale production. It offers an additional avenue for tailoring the properties of 2D boron materials and expanding their potential applications [41].

3.6 *Laser Ablation for 2D Boron Nanosheets*

By accurately aiming a high-powered laser beam onto a target made of bulk boron, laser ablation is a novel method for the synthesis of 2D boron nanosheets. Through the employment of a concentrated laser beam, boron atoms are locally heated and vaporized in the boron source material. Once on a substrate, these vaporized atoms condense and self-organize into 2D boron nanosheets like boron nanosheets or borophene. Behera and Behera, 2022 described that the capacity to create boron nanosheets with controlled thickness and form is one of the benefits of laser ablation. This technique also enables the tweaking of electrical characteristics by the insertion of additional components or dopants during the ablation process. It has the potential to expand the use of 2D boron nanosheets in electronics, sensing, and materials research since it is a flexible and accurate method for synthesis [42].

4 Comparison and Synergy Between Computational and Experimental Approaches in 2D Boron Nanosheets

According to Paul et al., 2017, computational and experimental techniques can work well together because they provide complementary information and capabilities for the study of 2D boron nanosheets. They also forecast the structural, electrical, and thermodynamic characteristics of fictitious boron nanosheets using computational simulations, notably DFT computations [43, 44].

Vaidyanathan et al., 2020 explained experimental methods that provide the vital validation and in-person confirmation of computer predictions. This allowed scientists to establish the presence of certain 2D boron nanosheets, ascertain their physical and chemical characteristics, and judge their applicability. Additionally, these experiments enabled the researchers to look into the elements based on temperature, pressure, and substrate interactions that may have an impact on the stability and behavior of these materials. Computational modeling and experimentation work have vast roles together to create a feedback loop that improves theoretical hypotheses and guides practical approaches, speeding up the creation and uses of 2D boron nanosheets in a variety of applications, from energy storage to nanoelectronics [45].

Additionally, these methods encourage creativity and problem-solving in the quickly developing area of 2D boron nanosheets. The difficulties brought on by the intricacy and reactivity of boron's structural makeup may be overcome by researchers by combining the advantages of computational and experimental approaches. In addition to improving our basic knowledge of boron-based materials, this multidisciplinary synergy holds the key to releasing their enormous potential for transformative technologies and applications, advancing materials science and nanotechnology [46].

5 Challenges and Opportunities

The field of 2D boron nanosheets presents a captivating blend of challenges and opportunities in materials science. One of the foremost challenges lies in the synthesis and stabilization of these inherently unstable structures due to their low-dimensional nature and high reactivity. Achieving controlled growth and scalable production methods remains a major hurdle. However, these challenges are met with an array of exciting opportunities. 2D boron nanosheets exhibit unique electronic, thermal, and mechanical properties that hold great promise for applications in next-generation electronics, energy storage, and catalysis. Exploring the fundamental properties and harnessing the remarkable characteristics of these materials could lead to groundbreaking technological advancements, revolutionizing various industries. Consequently, the pursuit of 2D boron nanosheets is not only a scientific adventure but also a gateway to transformative innovations in materials science and beyond.

6 Conclusion

In conclusion, the field of 2D boron nanosheets is a rapidly growing and fascinating frontier, offering a rich tapestry of challenges and opportunities. While the synthesis and stabilization of these materials present substantial hurdles, their unique properties, including exceptional thermal conductivity and electronic characteristics, open up a realm of possibilities across diverse fields, from electronics to clean energy solutions. The research on 2D boron nanosheets is still in its early stages, but the progress that has been made so far is encouraging. The synergy of both experimental and computational approaches plays a pivotal role in advancing our understanding of these materials. Experimental methods enable the synthesis and characterization of 2D boron nanosheets, providing crucial insights into their physical and chemical properties. Simultaneously, computational techniques, such as density functional theory and molecular dynamics simulations, empower us to explore their atomic-level behavior, predict novel structures, and comprehend their electronic properties. This collaborative approach leverages the strengths of both methodologies, accelerating our journey towards harnessing the full potential of 2D boron materials.

As we persist in unraveling the enigmatic nature of these nanosheets, the combined efforts of researchers from diverse disciplines promise to reshape industries and revolutionize our technological landscape. The synergy of experimental and computational methods, along with a commitment to scientific exploration, offers an exciting path forward, opening doors to groundbreaking innovations that could redefine the boundaries of materials science and technology. The collective pursuit of knowledge in this field embarks on a transformative journey with the potential to shape industries, create sustainable solutions, and drive technological progress for generations to come.

Acknowledgements The authors are thankful to the Drug Chemistry Research Centre, Indore, Madhya Pradesh, India, and the Director, CSIR-AMPRI, for their support and encouragement in this work.

Conflict of Interests The authors have declared no conflict of interest.

References

1. Gao, Z.D., Jiang, Z.H.Y., Li, J.D., Li, B.W., Long, Y.Y., Li, X.M., Yin, J., Guo, W.L.: Anisotropic mechanics of 2D materials. Adv. Eng. Mater. **24**(11), 2200519 (2022)
2. Zheng, Z., Huang, Y., Wu, F., Zhang, H., Fang, Z.: Multidimensional modulation of light fields via a combination of two-dimensional materials and meta-structures. Sci. China Inf. Sci. **66**(6), 160403 (2023)
3. Wang, W., Dong, Y., Xu, L., Dong, W., Niu, X., Lei, Z.: Combining bimetallic-alloy with selenium functionalized carbon to enhance electrocatalytic activity towards glucose oxidation. Electrochim. Acta **244**, 16–25 (2017)

4. Lin, Y., Connell, J.W.: Advances in 2D boron nitride nanosheets: nanosheets, nanoribbons, nanomeshes, and hybrids with graphene. Nanoscale **4**(22), 6908–6939 (2012)
5. Yadav, M., Dwibedi, V., Sharma, S., George, N.: Biogenic silica nanoparticles from agro-waste: Properties, mechanism of extraction and applications in environmental sustainability. J. Environ. Chem. Eng. 108550 (2022)
6. Kou, L., Ma, Y., Tang, C., Sun, Z., Du, A., Chen, C.: Auxetic and ferroelastic borophane: a novel 2D material with negative Possion's ratio and switchable dirac transport channels. Nano Lett. **16**(12), 7910–7914 (2016)
7. Lopez-Bezanilla, A., Littlewood, P.B.: Electronic properties of 8−Pmmn borophene. Phys. Rev. B **93**(24), 241405 (2016)
8. Wu, Z., Yin, Y., Hou, C., Tai, G.: Borophene reinforcing copper matrix composites: preparation and mechanical properties. J. Alloy. Compd. **930**, 167370 (2023)
9. Zhao, S., Wu, Y., Zhou, B., Liu, X.: Synthesis of boron nanosheets in copper medium. Sci. Rep. **9**(1), 17337 (2019)
10. Wang, X., Ji, S.L., Wang, X.Q., Bian, H.Y., Lin, L.R., Dai, H.Q., Xiao, H.: Thermally conductive, super flexible and flame-retardant BN-OH/PVA composite film reinforced by lignin nanoparticles. J. Mater. Chem. C **7**(45), 14159–14169 (2019)
11. Yadav, S., Sadique, M.A., Kaushik, A., Ranjan, P., Khan, R., Srivastava, A.K.: Borophene as an emerging 2D flatland for biomedical applications: current challenges and future prospects. J. Mater. Chem. B **10**(8), 1146–1175 (2022)
12. Shukla, M.K., Parihar, A., Karthikeyan, C., Kumar, D., Khan, R.: Multifunctional GQDs for receptor targeting, drug delivery, and bioimaging in pancreatic cancer. Nanoscale **15**(36), 14698–14716 (2023)
13. Sekkarapatti Ramasamy, M., Krishnamoorthi Kaliannagounder, V., Rahaman, A., Park, C.H., Kim, C.S., Kim, B.: Synergistic effect of reinforced multiwalled carbon nanotubes and boron nitride nanosheet-based hybrid piezoelectric PLLA scaffold for efficient bone tissue regeneration. ACS Biomater. Sci. Eng. **8**(8), 3542–3556 (2022)
14. Noreen, S., Tahir, M.B., Hussain, A., Nawaz, T., Rehman, J.U., Dahshan, A., Alzaid, M., Alrobei, H.: Emerging 2D-Nanostructured materials for electrochemical and sensing application-a review. Int. J. Hydrogen Energy **47**(2), 1371–1389 (2022)
15. Li, C., Tareen, A.K., Khan, K., Long, J., Hussain, I., Khan, M.F., Iqbal, M., Xie, Z., Zhang, Y., Mahmood, A., Mahmood, N.:. Highly efficient, remarkable sensor activity and energy storage properties of MXenes and Borophene nanomaterials. Prog. Solid State Chem. 100392 (2023)
16. Xu, R., Zou, X., Liu, B., Cheng, H.M.: Computational design and property predictions for two-dimensional nanosheets. Mater. Today **21**(4), 391–418 (2018)
17. Jana, D., Sun, C.L., Chen, L.C., Chen, K.H.: Effect of chemical doping of boron and nitrogen on the electronic, optical, and electrochemical properties of carbon nanotubes. Prog. Mater Sci. **58**(5), 565–635 (2013)
18. Bhimanapati, G.R., Lin, Z., Meunier, V., Jung, Y., Cha, J., Das, S., Xiao, D., Son, Y., Strano, M.S., Cooper, V.R., Liang, L.: Recent advances in two-dimensional materials beyond graphene. ACS Nano **9**(12), 11509–11539 (2015)
19. Naguib, M., Mochalin, V.N., Barsoum, M.W., Gogotsi, Y.: 25th anniversary article: MXenes: a new family of two-dimensional materials. Adv. Mater. **26**(7), 992–1005 (2014)
20. Kamel, M., Raissi, H., Hashemzadeh, H., Mohammadifard, K.: Theoretical elucidation of the amino acid interaction with graphene and functionalized graphene nanosheets: insights from DFT calculation and MD simulation. Amino Acids **52**, 1465–1478 (2020)
21. Timothy, R.A., Louis, H., Adindu, E.A., Gber, T.E., Agwamba, E.C., Offiong, O.E. Pembere, A.M.: Elucidation of collagen amino acid interactions with metals (B, Ni) encapsulated graphene/PEDOT material: Insight from DFT calculations and MD simulation. J. Mol. Liq. 122950 (2023)
22. Tang, X., Sun, W., Gu, Y., Lu, C., Kou, L., Chen, C.: CoB 6 monolayer: a robust two-dimensional ferromagnet. Phys. Rev. B **99**(4), 045445 (2019)
23. Fan, H., Luo, W., Dou, S., Zheng, Z.: Advanced two-dimensional materials toward polysulfides regulation of metal–sulfur batteries. SmartMat e1186 (2023)

24. Luo, W.W., Liu, G., Xu, Z.H., Zhou, Z.Y., Wang, X., Ouyang, C.Y., Liu, S.Q.: Insights into the physics of interaction between borophene and O2-first-principles investigation. Comput. Mater. Sci. **1**(140), 261–266 (2017)
25. Khan, K., Tareen, A.K., Aslam, M., Zhang, Y., Wang, R., Ouyang, Z., Zhang, H.: Recent advances in two-dimensional materials and their nanocomposites in sustainable energy conversion applications. Nanoscale **11**(45), 21622–21678 (2019)
26. Giustino, F., Lee, J.H., Trier, F., Bibes, M., Winter, S.M., Valentí, R., Son, Y.W., Taillefer, L., Heil, C., Figueroa, A.I., Plaçais, B.: The 2021 quantum materials roadmap. J. Phys.: Mater. **3**(4), 042006 (2021)
27. Amirian, B., Jafarzadeh, H., Abali, B.E., Reali, A., Hogan, J.D.: Thermodynamically-consistent derivation and computation of twinning and fracture in brittle materials by means of phase-field approaches in the finite element method. Int. J. Solids Struct. **252**, 111789 (2022)
28. Yin, H., Sun, Z., Wang, Z., Tang, D., Pang, C.H., Yu, X., Barnard, A.S., Zhao, H., Yin, Z.: The data-intensive scientific revolution occurring where two-dimensional materials meet machine learning. Cell Rep. Phys. Sci. **2**(7) (2021)
29. Shukla, M.K., Dubey, A., Pandey, S., Singh, S.K., Gupta, G., Prasher, P., Chellappan, D.K., Oliver, B.G., Kumar, D., Dua, K.: Managing apoptosis in lung diseases using nano-assisted drug delivery system. Curr. Pharm. Des. **28**(39), 3202–3211 (2022)
30. Suwardi, A., Wang, F., Xue, K., Han, M.Y., Teo, P., Wang, P., Wang, S., Liu, Y., Ye, E., Li, Z., Loh, X.J.: Machine learning-driven biomaterials evolution. Adv. Mater. **34**(1), 2102703 (2022)
31. Deokar, G., Jin, J., Schwingenschlögl, U., Costa, P.M.: Chemical vapor deposition-grown nitrogen-doped graphene's synthesis, characterization and applications. npj 2D Mater. Appl. **6**(1), 14 (2022)
32. Gautam, C., Chelliah, S.: Methods of hexagonal boron nitride exfoliation and its functionalization: covalent and non-covalent approaches. RSC Adv. **11**(50), 31284–31327 (2021)
33. Sozen, Y., Riquelme, J.J., Xie, Y., Munuera, C., Castellanos-Gomez, A.: High-throughput mechanical exfoliation for low-cost production of van der Waals nanosheets. Small Methods **15**, 2300326 (2023)
34. Wood, G.F., Zvoriste-Walters, C.E., Munday, M.G., Newton, M.E., Shkirskiy, V., Unwin, P.R., Macpherson, J.V.: High pressure high temperature synthesis of highly boron doped diamond microparticles and porous electrodes for electrochemical applications. Carbon **171**, 845–856 (2021)
35. Jin, M., Zhang, Y., Yan, C., Fu, Y., Guo, Y., Ma, X.: High-performance ionic liquid-based gel polymer electrolyte incorporating anion-trapping boron sites for all-solid-state supercapacitor application. ACS Appl. Mater. Interfaces. **10**(46), 39570–39580 (2018)
36. Yang, M., Jin, H., Sun, Z., Gui, R.: Experimental synthesis, functionalized modifications and potential applications of monoelemental zero-dimensional boron nanomaterials. J. Mater. Chem. A **10**(10), 5111–5146 (2022)
37. Rafiq, M., Hu, X., Ye, Z., Qayum, A., Xia, H., Hu, L., Lu, F., Chu, P.K.: Recent advances in structural engineering of 2D hexagonal boron nitride electrocatalysts. Nano Energy **91**, 106661 (2022)
38. Chakrabarty, S., Thakur, A., Rasyotra, A., Gaykwad, B., Jasuja, K.: Quasi-two-dimensional nanosheets from AlB2-type metal borides: physicochemical insights and emerging trends. J. Phys. Chem. C **127**(2), 852–870 (2022)
39. Zhang, K., Feng, Y., Wang, F., Yang, Z., Wang, J.: Two dimensional hexagonal boron nitride (2D-hBN): synthesis, properties and applications. J. Phys. Chem. C **5**(46), 11992–12022 (2017)
40. Niu, L., Coleman, J.N., Zhang, H., Shin, H., Chhowalla, M., Zheng, Z.: Production of two-dimensional nanomaterials via liquid-based direct exfoliation. Small **12**(3), 272–293 (2016)
41. Li, H., Jing, L., Liu, W., Lin, J., Tay, R.Y., Tsang, S.H., Teo, E.H.: Scalable production of few-layer boron sheets by liquid-phase exfoliation and their superior supercapacitive performance. ACS Nano **12**(2), 1262–1272 (2018)
42. Behera, A., Behera, A.: Nanomaterials. Adv. Mater.: Introd. Mod. Mater. Sci. 77–125 (2022)
43. Paul, J.T., Singh, A.K., Dong, Z., Zhuang, H., Revard, B.C., Rijal, B., Ashton, M., Linscheid, A., Blonsky, M., Gluhovic, D., Guo, J.: Computational methods for 2D materials: discovery,

property characterization, and application design. J. Phys.: Condens. Matter **29**(47), 473001 (2017)
44. Oganov, A.R., Pickard, C.J., Zhu, Q., Needs, R.J.: Structure prediction drives materials discovery. Nat. Rev. Mater. **4**(5), 331–348 (2019)
45. Vaidyanathan, A., Mathew, M., Radhakrishnan, S., Rout, C.S., Chakraborty, B.: Theoretical insight on the biosensing applications of 2D materials. J. Phys. Chem. B **124**(49), 11098–11122 (2020)
46. Zhao, Y., Waterhouse, G.I., Chen, G., Xiong, X., Wu, L.Z., Tung, C.H., Zhang, T.: Two-dimensional-related catalytic materials for solar-driven conversion of CO x into valuable chemical feedstocks. Chem. Soc. Rev. **48**(7), 1972–2010 (2019)

Properties of Boron Nanocrystals, Borophene, and its Polymorphs

Pushpesh Ranjan, Rudramani Tiwari, Raju Khan, Archana Singh, and Lal Singh Banjara

Abstract Boron and its polymorphs show a remarkable ability to form stable compounds due to their flexible and strong bonding capacity. They form stable multicenter two-electron bonds and two-center two-electron covalent bonds. They have been formed flexible two-dimensional (2D) materials that differ significantly from other well-studied 2D materials. Due to the remarkable properties of the 2D boron nanocrystals, and nanosheets, significant interest has been gaining in the recent past year. One of the great achievements of the syntheses of 2D boron sheets is known as borophene. It has gained notable attention due to its unique physical and chemical properties and is hence utilized in broad applications including electrochemical and optical sensors and biosensors, piezoelectric devices, catalysis reactions, energy storage devices, etc. In addition to this, boron also exists in several complex forms such as boron carbide (B_4C), boron nitride (BN) i.e. hexagonal boron nitride (hBN), rhombohedral boron nitride (rBN), zinc-blende boron nitride (cBN), and wurtzite boron nitride (wBN) which shown the unique physical and chemical properties. In this chapter, we have discussed a brief overview of the structural characteristics and different properties of boron nanocrystals, borophene, and other 2D boron polymorphs.

Keywords Boron nanocrystals · Borophene · 2D nanosheet · Physical · Chemical · Optical · Electronic properties

---

P. Ranjan · R. Khan (✉) · A. Singh
CSIR—Advanced Materials and Processes Research Institute (AMPRI), Bhopal 462026, India
e-mail: khan.raju@gmail.com

Academy of Scientific and Innovative Research (AcSIR), Ghaziabad 201002, India

R. Tiwari
Department of Chemistry, Institute of Science, Banaras Hindu University, Varanasi 221005, India

Regional Ayurveda Research Institute, Amkho, Gwalior 474009, India

L. S. Banjara
Govt Indira Gandhi Home Science Girls PG College, Shahdol 484001, India

R. Khan et al. (eds.), *2D Boron Nanosheets*, Engineering Materials,
https://doi.org/10.1007/978-981-97-7014-4_5

1 Introduction

Boron occupies a distinctive position as the sole non-metallic element in Group III of the periodic table. It exhibits exceptional chemical and physical properties, including low density, high melting points, bulk modulus, and hardness, and a notable neutron capture capacity [1]. Chemically boron is electron deficient and characterized by having three valence electrons available for bonding but four orbitals to be occupied. This electron deficiency enables boron to form distinctive bonding configurations, including two-center-two-electron (2c-2e) bonds and specialized multi-center-two-electron bonds [2, 3]. There are various bulk allotropes, with 16 distinct forms identified. Among these 16 forms; four of these allotropes (i.e. α-rhombohedral, β-rhombohedral, γ-orthorhombic, and β-tetragonal crystals) (Fig. 1a–g) are identified as thermodynamically stable and referred to as boron-icosahedral cluster solids (B-ICSs) [1, 4, 5]. B-ICSs are comprised of icosahedral close-cluster B_{12} units that link together or with other clusters to form diverse boron-rich compounds as polymorphs, including boron nitride (BN), boron arsenide ($B_{12}As_2$), boron sub-phosphide ($B_{12}P_2$), boron suboxide ($B_{12}O_2$), yttrium borides (YB_{66}), aluminum dodecaboride (AlB_{12}), and boron carbide (B_4C) [6–9]. In addition to 3D boron crystals, 2D structures exhibit remarkable diversity owing to their electron deficiency. 2D structures can be categorized into three main groups: graphene-like atomically monolayered boron sheets, 2D boron structures with a thickness of a single or a few unit cells, and other 2D boron structures [10]. It is significantly challenging to synthesize 2D boron due to the absence of layered bulk boron. However, recent advancements have seen the experimental synthesis of three types of monolayered boron sheets, known as borophenes, on an Ag substrate [11]. Also, the fabrication of few-layered 2D icosahedral boron sheets has been reported, stimulating considerable interest within the boron chemistry and 2D material communities [12].

Both 2D and 3D boron sheets exhibit notable resistance to oxidation attributed to their high planarity, stabilizing electronic states akin to graphene's electronic conjugation. While exposure to air leads to detectable oxygen content, oxidation primarily occurs at reactive edge states, leaving regions with perfect lattices mostly intact [13]. Moreover, Borophenes, lack the chemical stability same as graphene, readily contaminating upon extended air exposure [14]. This reactivity renders boron sheets potential catalysts for hydrogen evolution reactions (HER), unprecedented in other 2D materials. Functionalizing various chemical species, such as hydrogen, may yield boron hydrides with Dirac fermions [15]. Developing air-stable borophene-based devices necessitates packing techniques like silicon/silicon oxide capping layers to impede oxidation. Boron sheets exhibit a high capability for accommodating point defects, with low formation energies attributed to efficient passivation of dangling bonds through defect-induced atomic reconstruction, facilitated by delocalized multi-center bonds [13]. These are composed of classical covalent and metallic-like multi-center bonds and exhibit unique mechanical properties. Covalent bonds contribute to significant in-plane stiffness, while multi-center bonds impart fluxionality. In-plane stiffness is minimally affected by sheet curvature. Out-of-plane

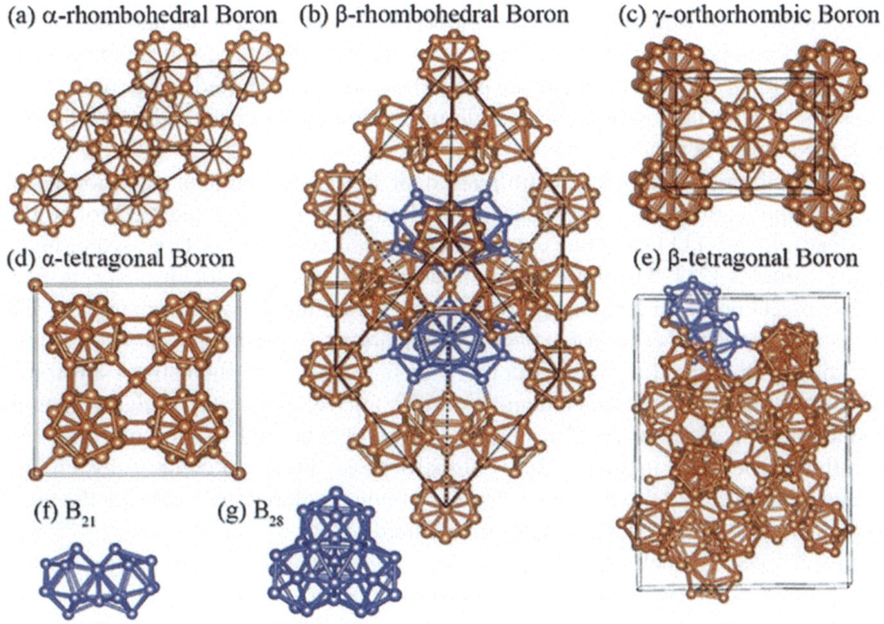

Fig. 1 The structures of boron allotropes: bulk polymorphs of **a** α-rhombohedral boron, **b** β-rhombohedral boron, **c** γ-orthorhombic boron, **d** α-tetragonal boron and **e** β-tetragonal boron, **f** structural units of B_{21} double-fused icosahedron and **g** B_{28} triply-fused icosahedron boron. Reprint with permission from Refs. [1, 23]

flexibility is pronounced, with lower bending modulus values, indicating potential as reinforcing elements in composites [16]. Borophenes show ideal strengths close to MoS_2's, with critical strains lower than graphene's, leading to structural phase transitions rather than failure under strain. Increasing tensile strain prompts structural reorganization, with increased hexagonal hole concentration relieving tension. These hexagonal holes are crucial for ductile breaking; sheets lacking them exhibit brittleness under stretching, emphasizing their importance in borophene mechanics [17]. The chemical bonding arrangements in both 2D and 3D boron structures share similarities, yet all known 2D boron polymorphs exhibit metallic behavior, in contrast to the insulating nature of their 3D counterparts. The metallic states primarily arise from the highly delocalized $2p_z$ state near the Fermi level, with contributions from the $2p_x$ and $2p_y$ states in certain cases [18]. Experimental verification of the predicted metallicity was achieved through angle-resolved photoemission spectroscopy (ARPES) measurements, confirming the presence of metallic bands derived from boron. Boron's lightweight suggests a potential for strong electron–phonon coupling, possibly leading to conventional superconductivity [19]. First-principles calculations support this notion, estimating critical temperatures (Tc) for superconductivity in the range of 10–20 K, higher than those reported for Li-decorated graphene. Phonon instabilities in free-standing polymorphs indicate potential for superconductivity, although Tc may be reduced by stronger substrate adhesion. Over the other boron polymorphs, Borophene offers possibilities for constructing thicker boron structures with novel electronic properties, such as distorted and double Dirac cones, as well as antiferromagnetic ordering, although experimental validation of these predictions is pending [20].

Despite these properties, the utilization of borophene and its polymorphs in practical applications encounters several challenges, primarily centered around the transfer from Ag(111) substrates and the controlled synthesis of large-scale, high-quality samples [21]. Theoretical investigations are essential for discovering new models of 2D boron in diverse environments, elucidating growth mechanisms, and uncovering novel properties for expanded applications. Experimental analysis should focus on developing efficient transfer methods to remove borophene from substrates and enhancing chemical vapor deposition (CVD) techniques for scalable sample production. The demonstrated properties of borophene and its polymorphs, and potential applications including 2D heterostructures, advanced composites, energy storage materials, and flexible electronics are being explored, highlighting the need for further research and development in this field [22].

This chapter has included a brief overview of the physical, chemical, and structural properties of the boron compounds such as boron carbide, boron nitride, borophene, boron oxides, boron quantum dots, and boron hydrides. Nevertheless, their extensive bondings, reactivity, stability, and ability to form various compounds including adducts have been discussed. Moreover, the massive applications of the boron compounds are also included.

2 Boron Materials and Their Properties

Boron is an amorphous powder that has a physical appearance is dark brown and/or black. It exhibits unique two-center two-electron covalent bonds and three-center two-electron bonds (banana bond in the case of diborane) because of the electron deficiency of the valence shell. These unique properties of the boron extended their compound formations in distinctive polymorphs. Nevertheless, pure boron is a crystalline and semiconductor property that has a black color appearance [24–28]. Due to semiconducting properties, it conducts electricity. However, its conductivity can be improved by adding impurities, doping heteroatoms, and/or by modifying its crystalline structure either through thermal treatment or mechanical methods. In addition to this, it possesses high strength, hardness, and roughness. Boron is fairly brittle, especially in its pure form. However, when combined with other elements to form compounds like boron carbide or borosilicate glass, its brittleness can be reduced. Moreover, one of the valuable properties of boron nanoparticles is their high-speed cutting ability. Their high hardness and strength make them effective tools for cutting and abrasion applications where they have been used for scratching a few abrasive materials like carborundum [13, 24, 29]. Also, Boron has a relatively high thermal conductivity of 27W/mK, which makes it a highly efficient material for heat transfer applications such as heat transfer in some ceramics and composite materials, and miniaturized devices [30].

Nonetheless, the granular sizes of boron nanoparticles vary, indicating that their shape and purities depend on the required applications. Because of this, there are several applications of the boron nanoparticles have been explored. Few boron compounds such as borosilicate glass (Pyrex) are transparent to electromagnetic radiation in the visible spectrum region. Also, boron-containing materials are advantageous in optical applications because of their remarkable transparent property. Due to these outstanding properties, boron and its composites are widely employed in many applications including ignition (because of its superior thermal characteristics). Also, the boron nanoparticles are utilized in fuels, torches, and charcoal briquettes. Nonetheless, they are employed in anti-freeze and corrosion inhibitor applications. In addition to this, their application also extended to photoelectric applications including fuel cells, electrolytic condensation, tanning, and other photoelectric sectors [26, 31–33].

Numerous polymorphs of boron exist in different structural forms based on their synthesis conditions. Some of the noteworthy polymorphs or allotropes of the boron materials are alpha (α)-boron, beta (β)-boron, boron nitride (BN), boron carbide (B_4C), borophene, boron oxides, boron hydride, borates, etc., where α-boron and β-boron are elemental forms of boron. α-Boron is also known as rhombohedral boron and it is the most stable form of elemental boron at room temperature and pressure. The high stability of α-boron due to the B_{12} icosahedra stacked in a rhombohedral lattice structure (Fig. 2) [1, 9, 18, 23, 32, 34, 35]. Its fascinating properties such as high strength and hardness are a result of this remarkable arrangement, which makes it valuable in a variety of applications, including abrasives and cutting tools. Apart from this, the β-Boron also known as cubic boron formed at high temperatures and pressures through CVD or shock compression processes. The β-Boron is metastable, which indicates that it is not the most stable form under normal environments, however, it can survive under specific conditions [26, 34, 36]. In contrast, β-boron shows unique properties due to its cubic form. Even though it is less common, β-Boron possesses unique qualities that make it useful in many applications, especially in nanotechnology and nanoscience and materials research. Its exceptional qualities and structure provide opportunities for new applications and the development of cutting-edge technology.

Apart from the elemental forms of boron, borate salts are also known as boric acid esters found in different forms based on the cations in the solution and the crystallization conditions. For instance, borax can crystallize into both hydrated and anhydrous forms. In many industrial applications such as ceramics, glassmaking, and detergent manufacture, where distinct crystal structures may display different characteristics, solubilities, or stabilities. The capacity of borates to exist in diverse polymorphs is also important. Understanding these polymorphic changes is essential to customizing borate materials for specific applications and maximizing their efficiency. Borates are used as an insecticide and a pest management tool [37, 38].

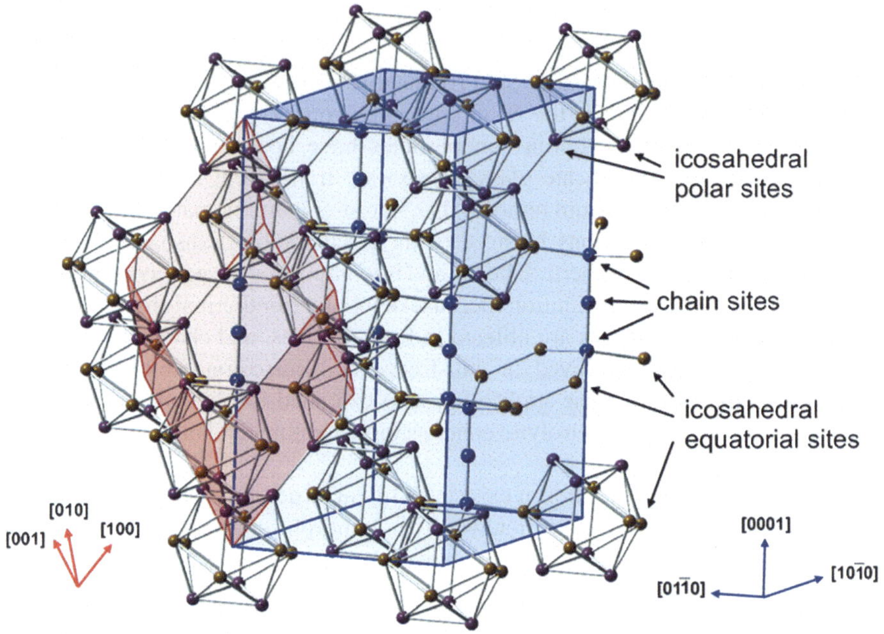

Fig. 2 The boron carbide lattice shows a correlation between the rhombohedral (red) and the hexagonal (blue) unit cells. Inequivalent lattice sites are marked by arrows. Reprint with permission from Ref. [9]

2.1 Boron Carbide

Another member of the boron family is boron carbide (B_4C), a ceramic material that exhibits a high elastic limit, high hardness, low density, superior electrical conductivity, and high melting point (>2400 °C). Also, it has excellent thermoelectric properties and can sustained at >700 °C [39, 40]. This ceramic material is employed for numerous applications due to its remarkable hardness and exceptional chemical inertness. Its complex structure consists of boron and carbon atoms arranged in a unique configuration, with icosahedral boron-rich clusters embedded within a carbon-rich matrix. This complex arrangement contributes to its outstanding properties and performance. Nevertheless, the composition of boron carbide can vary, leading to a range of properties depending on several factors such as the ratio of boron to carbon, the presence of impurities, and doped heteroatoms. This flexibility allows for the customization of boron carbide to enhance and improve its specific applications, whether it be in armor plating, cutting tools, abrasives, or nuclear shielding. Despite its exceptional hardness and toughness, boron carbide is a lightweight material, which makes it promising for desirable applications where weight reduction is an important factor such as in aerospace and military equipment. Additionally, its resistance to chemical corrosion and high temperatures sustainability further enhances its usefulness for harsh environments. They also exhibit exceptional semiconducting properties, where the low resistivity and low thermal conductivity offer them as a choice for the preparation of p-type semiconductor material which can be used at high temperatures for thermoelectric power conversion [26, 30].

Since boron carbide has p-type semiconductor properties, it has an abundance of positively charged "holes" that enhance electrical conductivity. When compared to other semiconductor ceramics such as silicon carbide (SiC), which usually has a band gap of about 3.0 eV, the significantly lower band gap of boron carbide was found to be around 1.64 eV. One of the reasons for the distinct electrical characteristics and semiconductor behavior of the boron carbide is its lower band gap. Compared to those materials that possess a comparatively higher band gap, boron carbide exhibits higher electrical conductivity because of its lower band gap which requires very low energy to transition the electrons from the valence band (VB) to the conduction band (CB). Hence, the exceptional semiconducting and electrical properties of boron carbide due to the low band gap extended their properties in electronic sensors, biosensors, and other semiconductor-based devices [41, 42].

It seems like the band gap of boron carbide can vary depending on the measurement method and the stoichiometry of the material. Werheit et al. reported an indirect band gap of 0.48 eV for the boron carbide through initial optical measurements, but in a later work, they reported a wider band gap of 2.09 eV. This suggests that boron carbide can exhibit a range of band gaps within the stoichiometric range of B_4C_3–$B_{11}C$ [9, 43]. Also, the theoretical band calculations dependably found larger band gaps, often more than 3.0 eV. This implies that current theoretical models may not be fully accountable for the disorder present in the material, which could lead to mid-gap states. The disparity between experimental and theoretical results underscores the complexity of the electronic structure of the boron carbide and the need for further research to better understand its properties [44, 45]. This is a direct result of the band theory, which states that, regardless of the technique of computation, a crystal with an odd number of valence electrons can be a metal. The Fermi level's location, which shows that (B_{12})CBC is metallic and $(B_{11}C)$CBC is semiconducting, is the primary distinction between the two structures. One characteristic of the band structure for the (B_{12})CCC configuration indicates the existence of a gap state of nonbonding nature, which is mostly attributed to the p orbital of the central carbon atom in the chain [9].

2.2 *Boron Nitride*

Boron nitride (BN) is another compound of the boron family. It is a ceramic material that possesses a high Hugoniot elastic limit, low density, high hardness, and remarkable electric conductivity [39, 46]. With the same composition of boron with nitrogen atoms, it displays isoelectronic and isostructural properties similar to carbon-based materials such as graphite, diamond, fullerene, nanotubes, and amorphous carbon. Boron nitride is found in both crystalline and amorphous forms. In crystalline form, it is found in hexagonal boron nitride (h-BN), sphalerite boron nitride (β-BN), and wurtzite boron nitride (ɣ-BN) (Fig. 3a). Apart from these, they are also found in a similar structural form as fullerenes nanotubes [26, 47–50]. The advantage of boron nitride is its strong oxidation resistance, which makes it suitable for use in high-temperature settings where carbon-based materials would deteriorate. Because, of the strong covalent bonds between boron and nitrogen (B–N) atoms which are less prone to oxidation than carbon–carbon (C–C) bonds (Fig. 3b) which cause high thermal stability [51–53]. Although most carbon compounds may be either conductive or semiconductive, however, boron nitride is an insulator. Hence this insulating property offers suitable applications such as for thermal and electrical insulator purposes. Nevertheless, while numerous carbon compounds have either indirect bandgaps or no bandgap at all, boron nitride possesses a direct bandgap. Because of its direct bandgap, boron nitride is a promising material for optoelectronic applications including photovoltaic cells and light-emitting diodes (LEDs). Furthermore, it also enables efficient light absorption and emission. Despite these merits, there are certain similarities between boron nitride and graphene, particularly concerning their hexagonal lattice structure. Because of these similarities, boron nitride is being widely investigated as a two-dimensional material having distinct mechanical and electrical characteristics, comparable to graphene [54–57].

Boron nitride has strong chemical and thermal stability, particularly h-BN and the c-BN cubic phase. At temperatures >1000 °C in air, >1400 °C in a vacuum, and up to 2850 °C in an inert environment, h-BN remains stable and does not break down [36, 58]. Also, h-BN has theoretical thermal conductivity values that are comparable to those of graphene, which is one of the best thermally conductive materials (~1700–2000 W/mK). At room temperature, the zigzag-edged boron nitride nanoribbon (BNNR) is 20% bigger than the armchair-edged nanoribbons, depending on the BNNR structure. At ambient temperature, the in-plane thermal conductivity of h-BN has been reported to reach 390 W/mK, which is almost 280 times higher

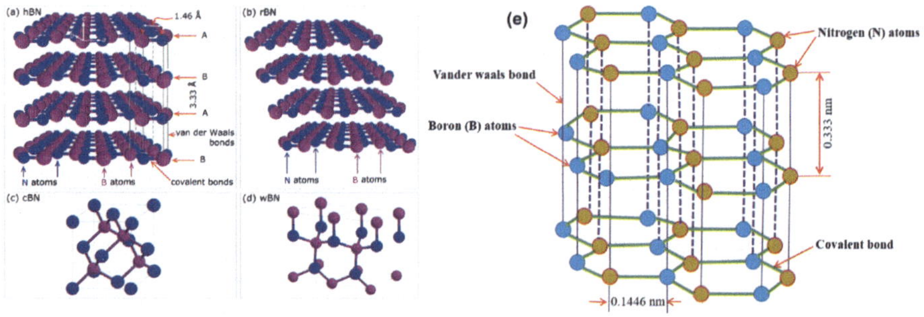

Fig. 3 Some crystal phases of boron nitride structures. **a** hexagonal boron nitride (hBN/αBN), **b** rhombohedral boron nitride (rBN), **c** cubic boron nitride (cBN/βBN), and **d** wurtzite boron nitride (wBN/γBN). Herein, B and N atoms are denoted as purple and blue balls, respectively. Reprint with permission from Ref. [50]. **e** Crystal structure representation of the hexagonal boron nitride (h–BN). Reprint with permission from Ref. [53]

than SiO_2. This makes h-BN an interesting dielectric material for electronic devices that produce heat. Also, because of its anisotropic bond strength, h-BN exhibits a significantly anisotropic coefficient of thermal expansion (CTE), much like graphite. At ambient temperature, the CTE in the c-direction is nearly 10 times greater and is 4.05×10^{-5} K^{-1} compared to the $-$ 2.90×10^{-6} K^{-1} in the a-direction (in-plane). The Weak van der Waals bonding between the planes is the cause of the significant positive thermal expansion in the c-direction [36, 47, 48, 59, 60].

The maximum lateral diameters (a few tens of μm) of the original starting material are frequently the limit of the boron nitride nanosheet (BNNS) that is produced by the exfoliation of these crystals. The structure of individual h-BN monolayers, also known as BNNSs, is a honeycomb structure and is made up of alternating nitrogen and boron atoms. For B–N, the binding type is covalent with a bond length of 1.45 Å. However, the centers of adjacent borazine rings are separated by 2.5 Å. Like graphene, the BNNS's edge structure can be zigzag or armchair-shaped. Whereas the latter is a boron nitride pair-edged structure, the former is a B or N-edged structure. Unlike graphite, the partly ionic structure of BN in h-BN decreases covalency and electrical conductivity. The electron-deficient B atoms are positioned just above or below the electron-rich N atoms in the next layer, which is often the most energetically advantageous stacking seen in BN. Over the several advantages, it has low strength, low toughness, and poor damage tolerance have limited their extensive usage as structural and functional materials [49, 61, 62].

A broad application for BN has been found in many scientific and engineering areas. This is mainly because of boron nitride, which has an intriguing characteristic known as "polymorphism," which is the capacity to crystallize into several different types of structures. Usually, this takes place by the variations in either the pressure or the temperature during the chemical reaction. Furthermore, even though the various polymorph structures, have the same chemical formula, their physical characteristics can be significantly changed. Because of this, polymorphs play a significant role in material design, and understanding how they improve the development of the task-specific polymorph is essential in this context [28, 49].

2.3 Borophene

2D monolayers of elemental boron, known as borophene, gain remarkable attention as promising materials with fascinating mechanical, chemical, optical, electrical, and catalytic properties. Moreover, it also has notable flexibility, low density, good elastic capability, and high thermal conductivity. In contrast to bulk boron allotropes, borophene has metallic characteristics consistent with predictions of a highly anisotropic two-dimensional metal. Since the bulk boron allotropes often behave as semiconductors or insulators. Because of borophene's distinct atomic arrangement and electronic structure, which provide a large density of electronic states at the Fermi level, this metallic characteristic develops [63–66]. It is found that the borophene exhibits highly anisotropic metallic behavior, which indicates that the direction of electron transport inside the material has a substantial impact on the conductivity of the material. The lattice structure of borophene and the orientation of its atomic bonds, which can be manipulated to modify its electronic characteristics, result in its anisotropic conductivity. Because borophene is extremely anisotropic and metallic, it has interesting prospects for use in electronics with specific conductivity and transport characteristics. Furthermore, borophene is a better choice for next nanoelectronic devices and quantum technologies due to its unique electronic structure and compatibility with two-dimensional materials. In addition to this, borophene exhibits a highly conductive property due to its metallic activity along specific crystallographic orientations. High-conductive materials are necessary for effective charge transport in devices like electrodes, transistors, and interconnects, hence this feature is helpful for electronic applications. Nevertheless, the borophene has negligible absorption or scattering of light due to its optical transparency along the "a" direction. This characteristic is essential for solar applications, where transparent conductive materials are required for electrodes to enable effective charge transfer and let light reach the active layers [67–70]. Besides, the remarkable anisotropic and mechanical properties, the pristine borophene displays high optical transparency due to the outstanding lesser thickness. This is because of the edge defect and localized three-dimensional atomic arrangements. The optical band gap of borophene was found to be <4 eV and it can vary from synthesis route since their structural arrangement can be altered [64, 66].

2.4 Boron Quantum Dots

Apart from the sheet structure, boron is also found in quantum dots. The ultra-violet (UV–Vis) absorption and photoluminescence (PL) spectroscopies were used to investigate the optical characteristics of boron quantum dots (B-QDs) and dodecaborane-surfactant complexes (DBS). The UV–vis spectra (Fig. 4a, b) for B-QDs suggest that they behave like semiconductors. This is determined from the absorption spectrum edge, which is indicative of semiconductors (The bandgap, which measures the least amount of energy needed for an electron to go from the valence band to the conduction band, indicates whether a material is semiconducting) and shows a rapid increase in absorption spectra at shorter wavelengths. Where the optical bandgap (Eg) of the B-QDs (Fig. 4a–f) was found to be 2.75 eV. It is assumed that the substantial quantum-confinement effect in the B-QDs is responsible for this value, which is significantly greater than that of the bulk boron (2.16 eV). The resulting 2.75 eV bandgap of boron quantum dots indicates that they may be a promising material for utilization in high-performance optoelectronic devices, such as solar. Because of its many unique anisotropy types, strong electrical conductivity along a direction, and optical transparency, B-QDs are a promising material with prospective applications in photovoltaics, flexible electronics, and display technologies [71, 72].

Researchers also reported that the B-QDs display colloidal characteristics. Where colloidal property of the B-QDs dispersion in an isopropyl alcohol solution is shown

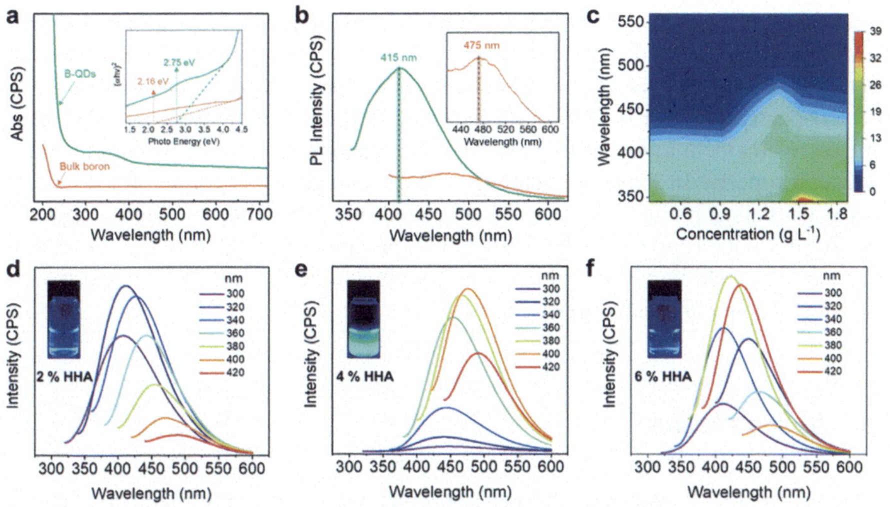

Fig. 4 Optical properties of B-QDs. **a** UV–vis spectra. **b** PL spectra excited by a 325 nm laser. **c** 3D mapping picture of B-QD concentration-dependent fluorescence excitation. **d–f** PL spectra of the DBS with different HHA concentrations excited at various wavelengths. The insets show the digital image under UV lamp irradiation of 365 nm. Reprint with permission from Ref. [71]

by the detection of the Tyndall effect. When light is dispersed by particles suspended in a liquid, it generates the Tyndall effect, which causes the light beam to become visible as it passes through the suspension. When the suspended particles are tiny and evenly distributed throughout the medium, as is the case with colloidal suspensions, these phenomena are more noticeable. The B-QDs may be finely distributed in the isopropyl alcohol, resulting in a colloidal suspension, as shown by their light black phase. A defining property of colloidal dispersions, the Tyndall effect shows visible proof that there are scattered particles in the fluid. Moreover, the B-QDs dispersion in isopropyl alcohol is stable, and its promise for practical applications. Since there is no discernible agglomeration or precipitation shown by the B-QDs over 60 days at ambient temperature. For a wide range of applications, this stability is essential since it guarantees constant qualities and performance throughout time. The B-QDs may also be well-suited for solution-based processes like inkjet printing, coating, or solution deposition, where uniform dispersion is necessary to achieve desired functionality and performance, as indicated by their lack of aggregation and precipitation. Stable colloidal dispersions are widely desired in a variety of sectors, including optoelectronics, sensors, and biomedical applications, and the boron quantum dots dispersion's continued stability underlines these applications' promise [73–76].

Two-photon fluorescence is also displayed by the B-QDs. When excited by a 360 nm laser, two emission peaks appear, resembling the one-photon fluorescence. When a fluorophore absorbs two photons with different energies at the same time, it produces two-photon fluorescence, which is the emission of one photon with a greater energy. In contrast to conventional one-photon excitation methods, this phenomenon enables the fluorophore to be excited by longer wavelength light, which lowers phototoxicity and for deeper tissue penetration. This implies that the B-QDs have special electrical characteristics that allow for effective two-photon absorption and emission. The prospective uses of B-QDs in bioimaging, fluorescence microscopy, and other optical sensing techniques are increased by their capacity to display this unique fluorescence property. In these applications, the advantages of two-photon excitation, such as decreased photobleaching and enhanced imaging depth, are highly appreciated. Additional research into the fundamental processes of B-QDs two-photon fluorescence may yield important information for enhancing the material's performance in a range of optical applications [71, 77, 78].

2.5 Boron Oxides

A class of chemical compounds known as boron oxides (Fig. 5a) is made up of atoms of both boron and oxygen. With the chemical formula B_2O_3, boric oxide is the most common type of boron oxide. It may take on several forms, such as amorphous, α-form, and β-form, depending on the synthesis conditions such as concentration, temperature, and pressure. There are further three crystalline forms of metaboric acid. Since orthoboric acid is constituted as $B_2O_3.3H_2O$ and metaboric acid as $B_2O_3.H_2O$, all of these forms of boric acid may be thought of as hydrates of boric oxide. Orthoboric acid's crystal structure is composed of planar sheets of triangular $B(OH)_3$ molecules joined by hydrogen bonds. The whole disarray in the molecular layer stacking pattern suggests that the van der Waals forces separating the planes are not very strong. This explains the crystals' easy cleaving into thin flakes and their slippery texture. Boron oxides are used in many different sectors and have a wide range of features. At room temperature, boronic oxide, also referred to as boron trioxide, is a white, crystalline solid. It is an essential substance used in the glass industry as a fluxing agent to reduce glass's melting point and increase its resilience to heat and chemicals. Additionally, borosilicate glass, which is frequently used in industrial applications, kitchenware, and laboratory glassware because of its exceptional thermal shock tolerance, is prepared using boric oxide. Boric oxide is used not just in glassmaking but also in the production of ceramic glazes, enamels, and flame retardants for a variety of materials [79–84].

Boric acid behaves like Lewis acid. It formed a four-coordinate borate complex and coordinated even with weak bases. It was experienced that the acidic strength increased by the strong electrolytes, where they rapidly formed a borate complex $[B(OH)_4^-]$. Moreover, the molten boric oxide is particularly corrosive to metals in the presence of oxygen because it dissolves most metal oxides at temperatures >1000 °C. Since, at temperatures exceeding 1000 °C, molten boric oxide easily combines with water vapor to create metaboric acid in the vapor state. It was also found that boric acids are soluble in water. Also, the solubility is influenced by the temperature, where temperature increases, boric acid becomes more soluble in water. Apart from this, the water solubility of boric acid may be increased or decreased by

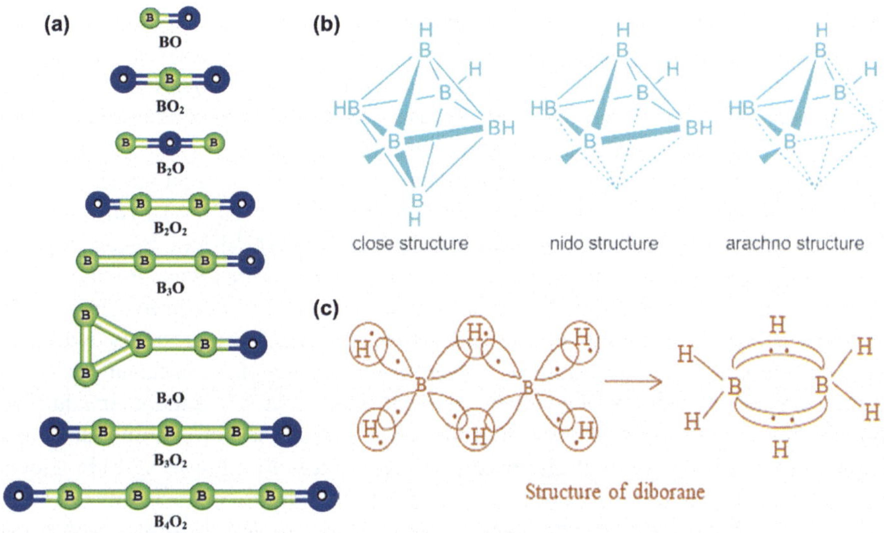

Fig. 5 **a** Shapes of the lowest-lying isomers of the neutral BnOm clusters with n) 1–4 and m) 1–2. **b** different structural of boron hydride compounds and **c** bonding in diborane. Reprint with permission from Ref. [82]

the addition of inorganic ions. Also, it is soluble in various organic solvents such as pyridine, dioxane, and diols and can form complexes with boric acid [21, 83, 84].

2.6 *Boron Hydrides*

Boron hydrides or boranes are a broad class of chemicals with a wide range of structures and characteristics. Diborane (B_2H_6), decaborane ($B_{10}H_{14}$), and closo-dodecaborate [$(B_{12}H_{12})^{2-}$] are a few examples (Fig. 5b). Diborane (B_2H_6) is the most stable basic form of boron hydride which exists in white gas at ambient temperature. Boron hydride materials have a broad variability of configurations, ranging from simple clusters to intricate polyhedral molecules. Diborane, which has two boron atoms and six hydrogen atoms organized in a bridged structure, is one well-known example of a three-centered two-electron bond (banana bond) (Fig. 5c). Due to its high energy content, diborane is well-known for its usage as a rocket fuel and as a reducing agent in organic synthesis. It may be found in several polymorphs, each with unique characteristics and structural configurations. Decaborane is used in many different processes, such as a dopant in semiconductor materials and as a precursor in the production of chemicals containing boron. Twelve boron and twelve hydrogen atoms in the closo-dodecaborate make them cage-like structures. Because of its capacity to aggregate specifically in tumor cells, it is employed in boron neutron capture therapy (BNCT), a treatment for certain tumors. Temperature, pressure, and chemical environment are some of the elements that might be responsible for the polymorphism seen in boranes. These structural changes result in a variety of features and possible uses in domains that span from materials science to medicine. Utilizing borane polymorphism to its fullest in a variety of technical and scientific efforts requires an understanding of and control over it [27, 85–88].

Boron hydride exhibits a very exothermic process due to the tremendous stability of the boron-oxygen bond and the relative instability of the boron hydrides. The dissociation enthalpy of the boron-oxygen bond is around 800 kJ/mol. Because boron hydrides are very unstable molecules, they can react in various ways involving the input of energy, such as thermal or electric discharge. Lower boron hydrides, such as B_2H_6 or B_4H_{10}, when pyrolyzed, undergo polymerization that yields hydrogen gas as a byproduct and a mixture of higher boron hydrides in significant amounts. Pyrolysis yields an amount of a yellow, solid substance that is flammable in addition to a combination of volatile boron hydrides. Also, since the boron compounds are acidic, they readily react with the base in its reaction with a boron hydride which allows single bonds to be produced at the expense of multi-center bonds, and the energy of this bond formation drives the reaction. The simplest examples of this are the cleavage reactions of bases with B_2H_6. Most bases except NH_3 cleave B_2H_6 in what is called a symmetrical manner. A significant amount of Lewis base adducts ($BH_3 \rightarrow NH_3$) or compounds has been formed when the Lewis base has attached itself to the boron hydride while maintaining the boron framework, are formed by boron hydrides [85, 86, 88].

3 Summary

The Boron, an electron-deficient atom possesses unique bonding capability to form various compounds. One of the remarkable bonding characteristics is two-center two-electron and multi-center two-electron bonds. The major class of boron compounds is boron carbide, boron nitride, borophene, boron quantum dots, boron oxides and hydrides, and other several polymorphs. They have exceptional physical and chemical properties including a two-dimensional sheet-like structure, low band gap, high strength, hardness, thermal stability, good electrical conductivity, and mechanical properties. Despite this, the Lewis acid characteristic and lower stability of some boron compounds readily react with bases to form a stable adduct. Despite their many nanostructures and associated characteristics, 2D boron crystals have demonstrated remarkable adaptability. Over the years, we believe 2D boron polymorphs will emerge as a special and indispensable material in nanotechnology and nanoscience.

References

1. Sun, X., Liu, X., Yin, J., et al.: Two-dimensional boron crystals: structural stability, tunable properties, fabrications and applications. Adv. Funct. Mater. **27**, 1603300 (2017). https://doi.org/10.1002/adfm.201603300
2. Jian, T., Chen, X., Li, S.-D., et al.: Probing the structures and bonding of size-selected boron and doped-boron clusters. Chem. Soc. Rev. **48**, 3550–3591 (2019). https://doi.org/10.1039/C9CS00233B
3. Wang, H., An, Q.: Band-gap engineering in high-temperature boron-rich icosahedral compounds. J. Phys. Chem. C **123**, 12505–12513 (2019). https://doi.org/10.1021/acs.jpcc.9b02254
4. Hayami, W., Hiroto, T., Soga, K., et al.: Thermodynamic stability of elemental boron allotropes with varying numbers of interstitial atoms. J. Solid State Chem. **329**, 124407 (2024). https://doi.org/10.1016/j.jssc.2023.124407
5. Wentorf, R.H.: Boron: another form. Science (1979) **147**, 49–50 (1965). https://doi.org/10.1126/science.147.3653.49
6. Yu, L.-F., Xu, J.-Y., Shen, C., et al.: Realizing ultra-low thermal conductivity by strong synergy of asymmetric geometry and electronic structure in boron nitride and arsenide. Rare Met. **42**, 210–221 (2023). https://doi.org/10.1007/s12598-022-02187-8
7. Slack, G.A., Morgan, K.E.: Some crystallography, chemistry, physics, and thermodynamics of B12O2, B12P2, B12As2, and related alpha-boron type crystals. J. Phys. Chem. Solids **75**, 1054–1074 (2014). https://doi.org/10.1016/j.jpcs.2013.12.021
8. Choi, S., Matsuo, J., Watanabe, T.: Synthesis of AlB12 and YB66 nanoparticles by RF thermal plasmas. J. Phys. Conf. Ser. **441**, 12030 (2013). https://doi.org/10.1088/1742-6596/441/1/012030
9. Domnich, V., Reynaud, S., Haber, R.A., Chhowalla, M.: Boron carbide: structure, properties, and stability under stress. J. Am. Ceram. Soc. **94**, 3605–3628 (2011). https://doi.org/10.1111/j.1551-2916.2011.04865.x
10. Fan, F.R., Wang, R., Zhang, H., Wu, W.: Emerging beyond-graphene elemental 2D materials for energy and catalysis applications. Chem. Soc. Rev. **50**, 10983–11031 (2021). https://doi.org/10.1039/C9CS00821G
11. Mannix, A.J., Zhou, X.-F., Kiraly, B., et al.: Synthesis of borophenes: anisotropic, two-dimensional boron polymorphs. Science (1979) **350**, 1513–1516 (2015). https://doi.org/10.1126/science.aad1080
12. Li, D., Gao, J., Cheng, P., et al.: 2D boron sheets: structure, growth, and electronic and thermal transport properties. Adv. Funct. Mater. **30**, 1904349 (2020). https://doi.org/10.1002/adfm.201904349
13. Zhang, Z., Penev, E.S., Yakobson, B.I.: Two-dimensional boron: structures, properties and applications. Chem. Soc. Rev. **46**, 6746–6763 (2017). https://doi.org/10.1039/C7CS00261K
14. Ou, M., Wang, X., Yu, L., et al.: The emergence and evolution of borophene. Adv. Sci. **8**, 2001801 (2021). https://doi.org/10.1002/advs.202001801
15. Jiao, Y., Ma, F., Bell, J., et al.: Two-dimensional boron hydride sheets: high stability, massless dirac fermions, and excellent mechanical properties. Angew. Chem. Int. Ed. **55**, 10292–10295 (2016). https://doi.org/10.1002/anie.201604369
16. Zhang, Z., Yang, Y., Penev, E.S., Yakobson, B.I.: Elasticity, flexibility, and ideal strength of borophenes. Adv. Funct. Mater. **27**, 1605059 (2017). https://doi.org/10.1002/adfm.201605059
17. Mozvashi, S.M., Mohebpour, M.A., Vishkayi, S.I., Tagani, M.B.: Mechanical strength and flexibility in α-4H borophene. Sci. Rep. **11**, 7547 (2021). https://doi.org/10.1038/s41598-021-87246-3
18. Penev, E.S., Bhowmick, S., Sadrzadeh, A., Yakobson, B.I.: Polymorphism of two-dimensional boron. Nano Lett. **12**, 2441–2445 (2012). https://doi.org/10.1021/nl3004754
19. Duan, Q., Zhan, L., Shen, J., et al.: Predicting superconductivity near 70 K in 1166-type boron-carbon clathrates at ambient pressure. Phys. Rev. B **109**, 54505 (2024). https://doi.org/10.1103/PhysRevB.109.054505

20. Krawiec, M.: Functionalization of group-14 two-dimensional materials. J. Phys.: Condens. Matter **30**, 233003 (2018). https://doi.org/10.1088/1361-648X/aac149
21. Shang, J., Ma, Y., Gu, Y., Kou, L.: Two dimensional boron nanosheets: synthesis, properties and applications. Phys. Chem. Chem. Phys. **20**, 28964–28978 (2018). https://doi.org/10.1039/C8CP04850A
22. Hou, C., Tai, G., Wu, Z., Hao, J.: Borophene: current status, challenges and opportunities. Chempluschem **85**, 2186–2196 (2020). https://doi.org/10.1002/cplu.202000550
23. Albert, B., Hillebrecht, H.: Boron: elementary challenge for experimenters and theoreticians. Angew. Chem. Int. Ed. **48**, 8640–8668 (2009). https://doi.org/10.1002/anie.200903246
24. Zhu, Y., Cai, J., Hosmane, N.S., Zhang, Y.: Introduction: basic concept of boron and its physical and chemical properties. In: Fundamentals and Applications of Boron Chemistry, pp. 1–57. Elsevier (2022). https://doi.org/10.1016/B978-0-12-822127-3.00003-X
25. Laubengayer, A.W., Hurd, D.T., Newkirk, A.E., Hoard, J.L.: Boron. I. Preparation and properties of pure crystalline boron. J. Am. Chem. Soc. **65**, 1924–1931 (1943). https://doi.org/10.1021/ja01250a036
26. Tatiya, S., Pandey, M., Bhattacharya, S.: Nanoparticles containing boron and its compounds—synthesis and applications: a review. J. Micromanufacturing **3**, 159–173 (2020). https://doi.org/10.1177/2516598420965319
27. DeFrancesco, H., Dudley, J., Coca, A.: Boron chemistry: an overview. In: Boron Reagents in Synthesis. American Chemical Society, p. 1 (2016)
28. Greenwood, N.N., Earnshaw, A.B.T.-C. of the E (Second E (1997) 6 - Boron. Butterworth-Heinemann, Oxford, pp. 139–215
29. Why is boron so hard? Nat. Mater. **9**, 6 (2010). https://doi.org/10.1038/nmat2600
30. Cai, Q., Scullion, D., Gan, W., et al.: High thermal conductivity of high-quality monolayer boron nitride and its thermal expansion. Sci. Adv. **5**, eaav0129 (2024). https://doi.org/10.1126/sciadv.aav0129
31. Gupta, G.H., Kadakia, S., Agiwal, D., et al.: Borophene nanomaterials: synthesis and applications in biosensors. Mater. Adv. **5**, 1803–1816 (2024). https://doi.org/10.1039/D3MA00829K
32. Zhou, S., Nozaki, T., Pi, X.: Boron nanocrystals as high-energy-density fuels. J. Phys. D Appl. Phys. **51**, 25305 (2018). https://doi.org/10.1088/1361-6463/aa9df6
33. Young, G., Sullivan, K., Zachariah, M.R., Yu, K.: Combustion characteristics of boron nanoparticles. Combust Flame **156**, 322–333 (2009). https://doi.org/10.1016/j.combustflame.2008.10.007
34. Ogitsu, T., Schwegler, E., Galli, G.: β-rhombohedral boron: at the crossroads of the chemistry of boron and the physics of frustration. Chem. Rev. **113**, 3425–3449 (2013). https://doi.org/10.1021/cr300356t
35. Fan, C., Li, J., Wang, L.: Phase transitions, mechanical properties and electronic structures of novel boron phases under high-pressure: a first-principles study. Sci. Rep. **4**, 6786 (2014). https://doi.org/10.1038/srep06786
36. Zhang, W., Chong, Y.M., He, B., Bello, I., Lee S.-T.: 3.24—Cubic boron nitride films: properties and applications. In: Sarin, V.K. (ed.) Comprehensive Hard Materials, pp. 607–639. Elsevier (2014), ISBN 9780080965284. https://doi.org/10.1016/B978-0-08-096527-7.00061-1.
37. Manankandayalage, C.P., Unruh, D.K., Krempner, C.: Boronic, diboronic and boric acid esters of 1,8-naphthalenediol—synthesis, structure and formation of boronium salts. Dalton Trans. **49**, 4834–4842 (2020). https://doi.org/10.1039/D0DT00745E
38. Bhagyaraj, S., Al-Ghouti, M.A., Kasak, P., Krupa, I.: An updated review on boron removal from water through adsorption processes. Emergent Mater **4**, 1167–1186 (2021). https://doi.org/10.1007/s42247-021-00197-3
39. Madhav Reddy, K., Guo, J.J., Shinoda, Y., et al.: Enhanced mechanical properties of nanocrystalline boron carbide by nanoporosity and interface phases. Nat. Commun. **3**, 1052 (2012). https://doi.org/10.1038/ncomms2047
40. Kozień, D., Jeleń, P., Stępień, J., et al.: Surface properties and morphology of boron carbide nanopowders obtained by lyophilization of saccharide precursors. Materials **14** (2021)

41. Werheit, H.: On excitons and other gap states in boron carbide. J. Phys.: Condens. Matter **18**, 10655 (2006). https://doi.org/10.1088/0953-8984/18/47/011
42. Cao, L.C.T., Hakim, L., Hsu, S.-H.: Boron doping in next-generation materials for semiconductor device. In: Wongchoosuk, C. (ed.) IntechOpen, Rijeka, p. Ch. 8 (2022)
43. Werheit, H., Binnenbruck, H., Hausen, A.: Optical properties of boron carbide and comparison with β-rhombohedral boron. Phys. Status Solidi (b) **47**,153–158 (1971). https://doi.org/10.1002/pssb.2220470118
44. Nosheen, U., Jalil, A., Ilyas, S.Z., et al.: First-principles insight into a B4C3 monolayer as a promising biosensor for exhaled breath analysis. J. Electron. Mater. **51**, 6568–6578 (2022). https://doi.org/10.1007/s11664-022-09898-9
45. Echeverria, E., Dong, B., Peterson, G., et al.: Semiconducting boron carbides with better charge extraction through the addition of pyridine moieties. J. Phys. D Appl. Phys. **49**, 355302 (2016). https://doi.org/10.1088/0022-3727/49/35/355302
46. Zhang, J., Tu, R., Goto, T.: 23—Cubic boron nitride-containing ceramic matrix composites for cutting tools. In: Advances in Ceramic Matrix Composites, pp. 570–586. Woodhead Publishing (2014). https://doi.org/10.1533/9780857098825.3.570
47. Biswas, A., Alvarez, G.A., Tripathi, M., et al.: Cubic and hexagonal boron nitride phases and phase boundaries. J. Mater. Chem. C Mater. **12**, 3053–3062 (2024). https://doi.org/10.1039/D4TC00039K
48. Bhimanapati, G.R., Glavin, N.R., Robinson, J.A.: 2D boron nitride: synthesis and applications. In: Semiconductors and Semimetals, pp. 101–147. Elsevier (2016). https://doi.org/10.1016/bs.semsem.2016.04.004
49. Merlo, A., Mokkapati, V.R.S.S., Pandit, S., Mijakovic, I.: Boron nitride nanomaterials: biocompatibility and bio-applications. Biomater. Sci. **6**, 2298–2311 (2018). https://doi.org/10.1039/C8BM00516H
50. Hoang, D.-Q., Vu, N.-H., Nguyen, T.-Q., et al.: Growth mechanisms of hBN crystalline nanostructures with rf sputtering deposition: challenges, opportunities, and future perspectives. Phys. Scr. **98**, 42001 (2023). https://doi.org/10.1088/1402-4896/acbe7b
51. Li, L.H., Cervenka, J., Watanabe, K., et al.: Strong oxidation resistance of atomically thin boron nitride nanosheets. ACS Nano **8**, 1457–1462 (2014). https://doi.org/10.1021/nn500059s
52. Angizi, S., Akbar, M.A., Darestani-Farahani, M., Kruse, P.: Review—two-dimensional boron carbon nitride: a comprehensive review. ECS J. Solid State Sci. Technol. **9**, 83004 (2020). https://doi.org/10.1149/2162-8777/abb8ef
53. Dadvand, M., Savadogo, O.: Effect of hBN on corrosion and wear performances of DC electrodeposited NiW and NiW–SiC on brass substrates. Coatings 12 (2022)
54. Jones, R.S., Gonzalez-Munoz, S., Griffiths, I., et al.: Thermal conductivity of carbon/boron nitride heteronanotube and boron nitride nanotube buckypapers: implications for thermal management composites. ACS Appl. Nano Mater. **6**, 15374–15384 (2023). https://doi.org/10.1021/acsanm.3c01147
55. Deb, J., Sarkar, U.: Boron-nitride and boron-phosphide doped twin-graphene: applications in electronics and optoelectronics. Appl. Surf. Sci. **541**, 148657 (2021). https://doi.org/10.1016/j.apsusc.2020.148657
56. Izyumskaya, N., Demchenko, D.O., Das, S., et al.: Recent development of boron nitride towards electronic applications. Adv. Electron. Mater. **3**, 1600485 (2017). https://doi.org/10.1002/aelm.201600485
57. Zhang, K., Feng, Y., Wang, F., et al.: Two dimensional hexagonal boron nitride (2D-hBN): synthesis, properties and applications. J. Mater. Chem. C Mater. **5**, 11992–12022 (2017). https://doi.org/10.1039/C7TC04300G
58. Oku, T.: Syntheses, Structures and Properties of Boron Nitride Nanoparticles BT - Handbook of Nanoparticles. In: International, S. (ed.) Aliofkhazraei M, pp. 1–32. Publishing, Cham (2015)
59. Lee, E.-S., Kang, J.-G., Kang, M.-K., et al.: High thermal conductivity of boron nitride filled epoxy composites prepared by tin solder nanoparticle decoration. Compos. B Eng. **225**, 109264 (2021). https://doi.org/10.1016/j.compositesb.2021.109264

60. Nguyen, H.T.T., Hanh, T.T.T.: Melting process of zigzag boron nitride nanoribbon. Physica E Low Dimens. Syst. Nanostruct. **106**, 95–100 (2019). https://doi.org/10.1016/j.physe.2018.10.029
61. Berski, S., Latajka, Z., Gordon, A.J.: On the multiple B-N bonding in boron compounds using the topological analysis of electron localization function (ELF). New J. Chem. **35**, 89–96 (2011). https://doi.org/10.1039/C0NJ00517G
62. Gadore, V., Mishra, S.R., Singh, A.K., Ahmaruzzaman, Md.: Advances in boron nitride-based nanomaterials for environmental remediation and water splitting: a review. RSC Adv. **14**, 3447–3472 (2024). https://doi.org/10.1039/D3RA08323C
63. Rahman, A., Rahman, M.T., Chowdhury, M.A., et al.: Emerging 2D borophene: synthesis, characterization, and sensing applications. Sens. Actuators A Phys. **359**, 114468 (2023). https://doi.org/10.1016/j.sna.2023.114468
64. Yadav, S., MohdA, S., Kaushik, A., et al.: Borophene as an emerging 2D flatland for biomedical applications: current challenges and future prospects. J. Mater. Chem. B **10**, 1146–1175 (2022). https://doi.org/10.1039/D1TB02277F
65. Kaneti, Y.V., Benu, D.P., Xu, X., et al.: Borophene: two-dimensional boron monolayer: synthesis, properties, and potential applications. Nano Lett. **122**, 6746–6763 (2012). https://doi.org/10.1021/nl3004754
66. Asaduzzaman Chowdhury, M., Kamal Uddin, M.M., Masud Rana, Md., et al.: Synthesis and characterization of borophene for future applications. Results Chem. **6**, 101071 (2023). https://doi.org/10.1016/j.rechem.2023.101071
67. Xie, Z., Meng, X., Li, X., et al.: Two-dimensional borophene: properties, fabrication, and promising applications. Research 2020 (2024). https://doi.org/10.34133/2020/2624617
68. Ayodhya, D., Veerabhadram, G.: A brief review on synthesis, properties and lithium-ion battery applications of borophene. FlatChem **19**, 100150 (2020). https://doi.org/10.1016/j.flatc.2019.100150
69. Jiang, H.R., Lu, Z., Wu, M.C., et al.: Borophene: a promising anode material offering high specific capacity and high rate capability for lithium-ion batteries. Nano Energy **23**, 97–104 (2016). https://doi.org/10.1016/j.nanoen.2016.03.013
70. Abasi, T., Boochani, A., Masharian, S.R.: Metallic and intra-band investigation of optical properties for Borophene nano-sheet: a DFT study. Int. Nano Lett. **10**, 33–41 (2020). https://doi.org/10.1007/s40089-019-00288-4
71. Wang, H., An, D., Wang, M., et al.: Crystalline borophene quantum dots and their derivative boron nanospheres. Mater. Adv. **2**, 3269–3273 (2021). https://doi.org/10.1039/D1MA00124H
72. Rawat, J., Sajwan, D., Garimella, S.V., et al.: Boron nitride quantum dots: a rising star in sensing applications. Nano Trends **2**, 100008 (2023). https://doi.org/10.1016/j.nwnano.2023.100008
73. Gogoi, D., Hazarika, C., Neog, G., et al.: Borophene quantum dots as novel peroxidase-mimicking nanozyme: a dual-mode assay for the detection of oxytetracycline and tetracycline antibiotics. ACS Appl. Mater. Interfaces **16**, 14645–14660 (2024). https://doi.org/10.1021/acsami.3c12108
74. Aswal, S., Ghosh, K., Giri, P.K.: Borophene quantum dots with strong photoluminescence for selective metal ion sensing. ACS Appl. Nano Mater. (2024). https://doi.org/10.1021/acsanm.4c01204
75. Wang, C., Chen, Q., Chen, H., et al.: Boron quantum dots all-optical modulator based on efficient photothermal effect. Opto-Electron. Adv. **4**, 200032 (2021). https://doi.org/10.29026/oea.2021.200032
76. Yang, M., Jin, H., Gui, R.: Ag+-doped boron quantum dots with enhanced stability and fluorescence enabling versatile practicality in visual detection, sensing, imaging and photocatalytic degradation. J. Colloid Interface Sci. **639**, 49–58 (2023). https://doi.org/10.1016/j.jcis.2023.02.066
77. Dehghani, A., Madadi Ardekani, S., Lesani, P., et al.: Two-photon active boron nitride quantum dots for multiplexed imaging, intracellular ferric ion biosensing, and pH tracking in living cells. ACS Appl. Bio Mater. **1**, 975–984 (2018). https://doi.org/10.1021/acsabm.8b00145

78. Acharya, A., Sharma, S., Liu, X., et al.: A review on van der Waals boron nitride quantum dots. C (Basel) **7** (2021)
79. Park, S.-J., Seo, M.-K.: Element and processing. In: Interface Science and Technology, pp. 431–499. Elsevier (2011). https://doi.org/10.1016/B978-0-12-375049-5.00006-2
80. Varshneya, A.K., Mauro, J.C.: Introduction. In: Fundamentals of inorganic glasses, 3rd edn., pp. 1–18. Elsevier (2019). https://doi.org/10.1016/B978-0-12-816225-5.00001-8
81. von Burg, R.: Toxicology update. J. Appl. Toxicol. **12**, 149–152 (1992). https://doi.org/10.1002/jat.2550120213
82. Tai, T.B., Nguyen, M.T., Dixon, D.A.: Thermochemical properties and electronic structure of boron oxides BnOm (n = 5–10, m = 1–2) and their anions. J. Phys. Chem. A **114**, 2893–2912 (2010). https://doi.org/10.1021/jp909512m
83. Nguyen, M.T., Matus, M.H., Ngan, V.T., et al.: Thermochemistry and electronic structure of small boron and boron oxide clusters and their anions. J. Phys. Chem. A **113**, 4895–4909 (2009). https://doi.org/10.1021/jp811391v
84. Briggs, M.: Boron oxides, boric acid, and borates. In: Kirk-Othmer Encyclopedia of Chemical Technology (2001)
85. Bauer, S.H.: Structures and physical properties of the hydrides of boron and of their derivatives. Chem. Rev. **31**, 43–75 (1942). https://doi.org/10.1021/cr60098a002
86. Siegel, B., Mack, J.L.: The boron hydrides. J. Chem. Educ. **34**, 314 (1957). https://doi.org/10.1021/ed034p314
87. Bregadze, I.V., Sivaev, B.I., Glazun, A.S.: Polyhedral boron compounds as potential diagnostic and therapeutic antitumor agents. Anticancer Agents Med Chem. **6**, 75–109 (2006)
88. Beall, H., Gaines, D.F.: Boron hydrides. In: Encyclopedia of Physical Science and Technology, 3rd edn., pp. 301–316. Academic Press, New York (2003). https://doi.org/10.1016/B0-12-227410-5/00073-9

Boron-Based Two-Dimensional Nanosheets in Energy Devices

Usama Ansari, Shivam Kashyap, Sumit Kumar, Sanjeeve Thakur, Deepak Poddar, and Ankita Singh

Abstract In the current scenario, 2D nanomaterials such as graphene, transition metal dichalcogenides, Mxenes, borophene, etc. play a significant role in the conversion of energy and its storage. Among these 2D nanomaterials, borophene shows versatile behavior because it consists of delocalized five-centered three-electron bonds, which leads to the metallic behavior of most borophene polymorphs. Borophene has become popular these days because the β_{12} phase of borophene delivers a considerable orbital overlap that gives a large electron density around the conduction level, which leads to a metallic behavior. The 2D boron nanosheet can be chemically functionalized by grafting, anchoring, etc., and exhibits remarkable properties like large surface area, high Young's modulus, notable high Fermi velocity, flexibility, and optical properties that allow borophene to be employed in electronic and energy devices such as organic light-emitting diodes (OLEDs), fuel cells, storage batteries, oxygen evolution reaction (OER), oxidation–reduction reaction, and activating energy-rich small molecules. In this chapter, the recent advancements in the different polymorphs of borophene and its composites for energy conversion and storage devices are discussed, focusing on batteries, catalysis, and fuel cells for hydrogen storage.

Keywords Boron nanosheets · Renewable energy · Chemical functionality · Batteries · Fuel cells

1 Introduction

This era is the golden period of 2D materials for energy applications and devices because they offer great optical, mechanical, magnetic, and electrical properties, which opens doors for researchers to use in the field of energy systems like storage, conversion, and fuel cell applications. 2D materials such as graphene, transition metal dichalcogenides (TMDCs), hexagonal boron nitride (h-BN), boron carbides

U. Ansari · S. Kashyap · S. Kumar · S. Thakur · D. Poddar · A. Singh (✉)
Department of Chemistry, Netaji Subhas University of Technology (N.S.U.T), Azad Hind Fauj Marg, Sector-3, Dwarka, New Delhi 110078, India

R. Khan et al. (eds.), *2D Boron Nanosheets*, Engineering Materials,
https://doi.org/10.1007/978-981-97-7014-4_6

(BC), and borophene are some interesting 2D materials that are frequently used in energy systems. Among all these, boron-based 2D materials have gained tremendous interest due to their bridging position between metals and non-metals, and their versatile behavior towards chemical reactions. For instance, from the perspective of electronic structure, the electron deficit nature of boron serves it to act as an electrophile or Lewis's acid however, in some circumstances, it can be polarized and negatively charged, where it behaves as a nucleophile or Lewis's base, like B–N bonded compounds. A wide range of bulky and asymmetrical borate anions and anionic boron clusters have also drawn interest due to their fundamentally low affinity for cations, making them vital parts of electrochemical devices [1].

Boron-based nanosheets (BBNS), specifically borophene, have a single layer of boron atoms similar to the carbon in graphene, which leads to its unique properties like low density, high mechanical strength, transport ability, and superconductivity, which are suitable for many technologies like energy, catalysis, and sensing applications [2, 3]. Borophene exists in various allotropes; among them, some allotropes are considered metallic, like β_{12}, X_3, honeycomb, and 2-Pmm [4]. As a less explored material as compared to other 2D materials, it possesses advantages such as high surface properties and an improved chemical structure [5]. Apart from this, due to boron's special ability to form three-center two-electron bonds, borophene has more e^-density around the conduction level, i.e., the fermi level shifts towards the conduction band, which is distinct from comparable materials like graphene, which have a fermi level in the middle between the valence and conduction bands [6]. Borophene is used for designing composites to use in flexible nanodevices due to its great flexibility against off-plane deformation. The magnetic and electronic properties of borophene can be easily controlled for energy and optoelectronics because of its anisotropic structure, and the low mass density of boron. There is also a substantial electron–phonon coupling interaction between 10,000 and 20,000 which leads to phonon-mediated superconductivity at high critical temperatures [2]. The electron mobility of a material depends on its properties like atomic size, charge, strain in the system, atomic bonding, and crystal structure. Thus, among all the existing 2D materials, graphene has the highest electron mobility, which occurs at around 180,000 cm^2-V^{-1}-s^{-1} [7]. Unlike this, its cousins like stanine, germanene, silicene, and phosphorene have moderate electronic mobility, and after that, some other materials with lower electron mobility are also present, like transition metal dichalcogenides. Notably, in a recent study, the electron mobility of borophene was reported to be 280,000 cm^2V^{-1} s^{-1} which created interest in this material for use in the field of energy storage and sensors [8].

Apart from borophene, there are some other boron-based 2D materials of interest in the field of energy devices, like BN, BC, boron phosphide, Boron oxide, etc. BN, which has boron and nitrogen atoms, may be found in many different forms, such as cubic BN (c-BN), wurtzite BN (w-BN), and hexagonal BN (h-BN). Among these forms of BN, h-BN shows more potential for use in energy devices due to its structural, mechanical, optical, and thermal properties. h-BN is known as white graphene because it has a graphene-like structure in which planar hexagonal networks are regularly stacked where carbon atoms are replaced by boron and nitrogen atoms, respectively [9]. Along with this, h-BN has a larger lattice constant than graphene, which provides it with high thermal conductivity and tribological properties that allow h-BN to be used as a lubricant or in high heat-resistant ceramic materials [10]. The electronegativity difference between nitrogen and boron in sp^2 hybridization with π-stacking contributes towards a wide bandgap, which is around 5.5 eV. However, the performance of h-BN will be optimized by adjusting the determinative factors like surface modification, crystallite size, dispersibility, and layer thickness. h-BN shows great resistivity of ~1018 Ω-cm, strong corrosion resistance, high oxidation resistance above 800 °C, great thermal conductance, and excellent thermal stability ~3000 °C under N_2 conditions with a low thermal expansion coefficient. These properties, combined with 2D materials properties, make h-BN a prominent candidate to use as a lubricant or effective additive. Utilizing BNNS insulating properties, BN is applied as a dielectric layer in the fields of electronics and energy. Additionally, BNNS serve as fillers in nanocomposites for dielectric, capacitive, and thermal energy applications. Recent developments in batteries and supercapacitors offer a platform for applications of BNNS, such as in the electrode interlayer, electrolyte additive, separator coating, etc. [11]. Fundamentally, the BNNS's chemical and physical characteristics, such as its crystalline structure, lateral size, layers, and surface functional groups, determine the combined properties and eventual performances of BNNS in electrical and energy devices. To produce superior BNNS and satisfy real-world demand, it is, therefore, necessary to comprehensively examine the synthesis techniques and functionalization approach.

In continuation of the above materials, BC is a ceramic substance with a high melting point, superior thermal conductivity, and great hardness that might be used in energy devices. One of the most common uses of BC is as a neutron absorber in nuclear reactors because BC absorbs neutrons with a large cross-section, making it helpful for regulating nuclear processes [12]. BC has been investigated for its possible application in energy storage devices such as batteries and supercapacitors, in addition to its usage in nuclear reactors. BC has a high hypothetic specific capacity for lithium-ion battery anodes, which implies it might theoretically store more energy than other Li-ion battery materials now in use. Furthermore, BC has been studied for its possible use in thermoelectric devices. BCs excellent thermal conductivity makes it a possible option for use in such devices. Overall, the unique structures of BC make it a viable material for use in energy devices and more studies might help unlock its full potential in these applications [13].

This chapter aims to investigate and present the current research approaches on synthesis strategies of 2D boron-based materials (BBM) such as borophene, h-BN, BC, etc. for energy applications, highlighting their unique properties and potential for energy conversion and storage devices. The graphical representation of the BBM and its potential applications is given in Fig. 1. After reading this book chapter, readers also get to know how these materials revolutionize the field of nanomaterials and renewable energy and pave the way for frontiers in 2D materials. Furthermore, this book chapter proposes that BBNS can be synthesized using a variety of

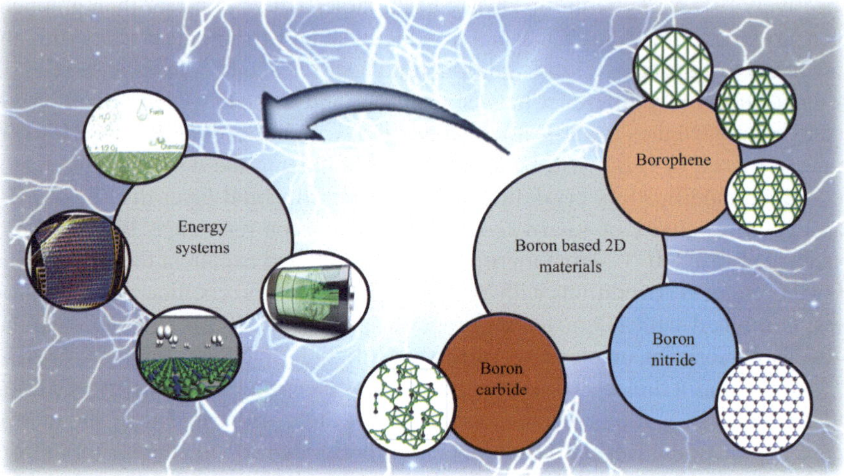

Fig. 1 Uses of boron-based materials in possible energy applications

novel methods, including chemical vapor deposition (CVD), physical vapor deposition (PVD), and solution-based approaches like liquid exfoliation, and their structural, mechanical, optical, magnetic, and electronic properties of BBNS will provide critical insights into their potential applications and enable the design of tailored materials for specific purposes.

2 Synthesis

The quality, size, layer thickness, and functional groups of BBNS depend greatly on the synthesis method. As a result, the first understanding of BBNS designs comes from the objectives that underpin the suggested applications. For example, the large size and crystalline BNNS produced by CVD may directly contribute to the atomic layer for 2D heterojunctions, in contrast to the solvent-exfoliated BNNS, which are often utilized as fillers for thermally conductive and dielectric polymer composites. Additionally, the second stage incorporates knowledge of physics and chemistry into the synthesis. Mechanical shearing, thermal expansion, crystallinity, covalent bonding, noncovalent bonding, etc. are some of these processes, which range in complexity from simple to complicated [2]. Additionally, a variety of approaches are being discussed and contrasted to produce the 2D materials which are represented in Fig. 2. Some of these approaches are relatively broad, while others call for particular circumstances. For electronics and energy applications, reliable methods for manufacturing BBNS with consistent properties are essentially required. Here, the benefits of bottom-up synthesis and top-down exfoliation approaches are examined and explained, respectively [14].

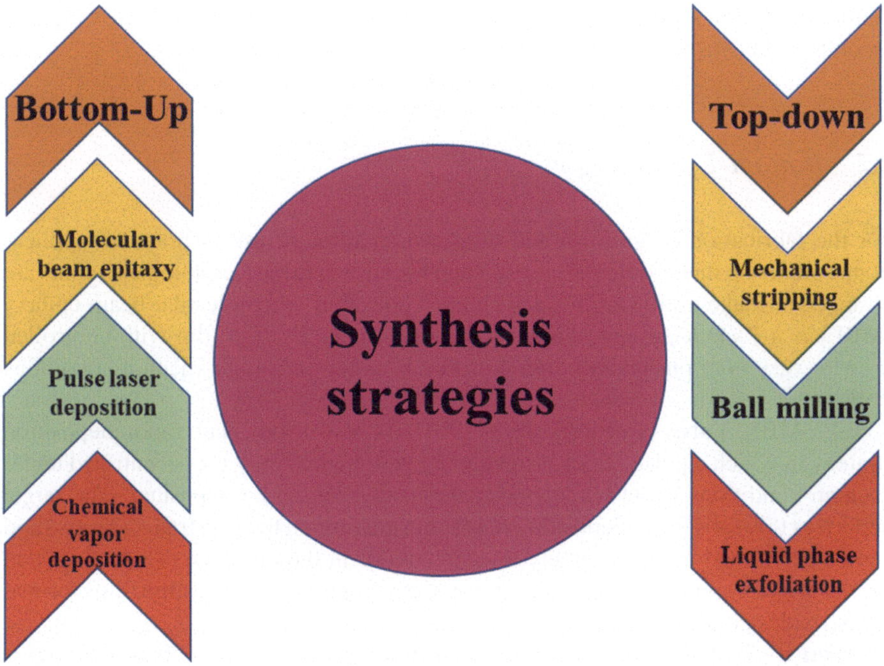

Fig. 2 synthesis methods of boron-based 2D materials which include majorly two strategies bottom-up and top-down

2.1 Top-Down Strategies

There are many fabrication technologies to fabricate thin layers from their bulk materials are known as top-down strategies in which mechanical striping or peeling is well known or the simplest technique, in this technique adhesive tape, is used to attach to a single crystal and peeled to transfer on other targeted substrate [15]. The fabricated layer of BBNS is seen and identified with the help of optical microscopy. This method is also effectively used for other 2D materials like TMDCs, graphene, etc. [16].

Ball milling exfoliation is a crucial top-down process that uses a planetary mill to mix and spin the balls, bulk materials, and milling agents to shear the thickness of 2D materials. The efficiency of this method is about 85% [17].

Liquid phase exfoliation is another powerful technique to achieve BBNSs. In this method, proper solvents are used to get nano sheets like distilled water, surfactant solution, and organic solvents are used, and the raw ingredients are often sonicated for many hours after being dissolved in an aqueous solution. The 2D crystallites are exfoliated into nanosheets having a few hundred nanometres thickness by cavitation action. After electrostatic repulsion, polymer solvents were employed to stabilize the colloidal solution after exfoliation. This method has a low production rate of nearly around 10% [18].

2.2 Bottom-Up Strategies

For the fabrication of uniform and high-quality films to use in practical applications bottom-up methodologies are adopted because in these technologies' atoms are arranged to make uniform films and can be controlled like molecular beam epitaxy (MBE) is a PVD technique that provides the atomically thin film with controlled growth, very high purity, and high uniformity under ultra-high vacuum conditions [19].

Like MBE, pulse laser deposition (PLD) is also a physical vapor deposition strategy in which a pulsed laser beam having very high energy is concentrated under vacuum conditions towards targeted material that has to be deposited. The target vaporizes this substance, depositing it as a thin coating on the substrate. This method can be performed in an extremely low vacuum or in the presence of a background gas, such as oxygen, which is commonly used to properly oxygenate the films formed while oxides are deposited [20].

Apart from PVD, in the chemical deposition process, the solid bulk material is vaporized and deposited on the substrate surface as a result of a chemical reaction [21]. CVD has different types according to the vacuum condition used while fabricating the film like atmospheric pressure CVD (APCVD), low-pressure CVD (LPCVD), and ultra-high vacuum CVD (UHCVD). Some examples of boron base materials fabricated from different techniques are shown in Table 1.

3 Properties

Depending upon the different types of synthesis methods, 2D boron sheets mainly borophene and BN show diversified properties which can be used in various applications. The properties are as follows.

3.1 Mechanical Properties

The basic mechanical properties of 2D flatlands are crucial for the manufacturing of new nanomaterials and next-generation nanocomposites [31]. It has been reported that borophene has an ultrahigh mechanical modulus due to its strong boron-boron bonds and atomic structure. There have been reports on the elastic constants, shear, layer, Young's modulus (N/m), and Poisson's ratio of 2D materials, such as

Table 1 Synthesis of boron-based materials via different methods to use in different applications

S.N	Boron based 2 D material	Fabrication Technique	Substrate	Application	References
1	Borophene	Thermal deposition	Au	As electrode	[22]
2	Borophene nanoribbon	MBE	Ag (110)	–	[23]
3	Borophene	CVD	Quartz	Hydroelectric generator	[24]
4	Few layer borophene	MBE	Ru (0001)	High-quality borophene on metal substrate	[25]
5	BN	Ball milling	-	BN aerogel and freestanding membranes	[26]
6	BNNS	Liquid phase exfoliation		Thin film	[27]
7	h-BN	APCVD	Cu	Nanoscale optics and photonics	[28]
8	h-BN	APCVD	Ni	Thin film	[29]
9	h-BN	LPCVD	Pt	Single layer	[30]
10	h-BN	Mechanical exfoliation	Sio_2/Si	Optoelectronics	[3]

borophene, BN, and BC [32]. Borophene is an extremely strong and flexible material with Young's modulus that is comparable to graphene. Such properties make them an ideal candidate for lightweight, high-strength material applications. There is a strong correlation between Borophene's crystal structure and its mechanical properties. There are several types of crystal structures for borophene, each with a different mechanical property. The elastic and mechanical features of BC are derived from atomic bonding abilities such as localization and delocalization, ionicity and covalency of the bonds, and electron density in inter-atomic regions. More localized covalent connections and higher interatomic electron density are specifically linked to greater stiffness and hardness [33]. The h-BN material is highly tensile and mechanically robust. In addition to its strength, it is more resistant to thermal degradation and oxidation than graphene. Due to this property, it can be used in a variety of structural applications [34].

3.2 Optical Properties

2D nanosheets that are made of boron have distinctive optical characteristics which distinguish them from graphene. The high absorption coefficient of borophene is due to its electronic band structure, making it excellent for use in solar and photovoltaic applications. Particularly in the ultraviolet and visible portion of the electromagnetic spectrum, the anisotropy of the electronic band structure results in a high absorption

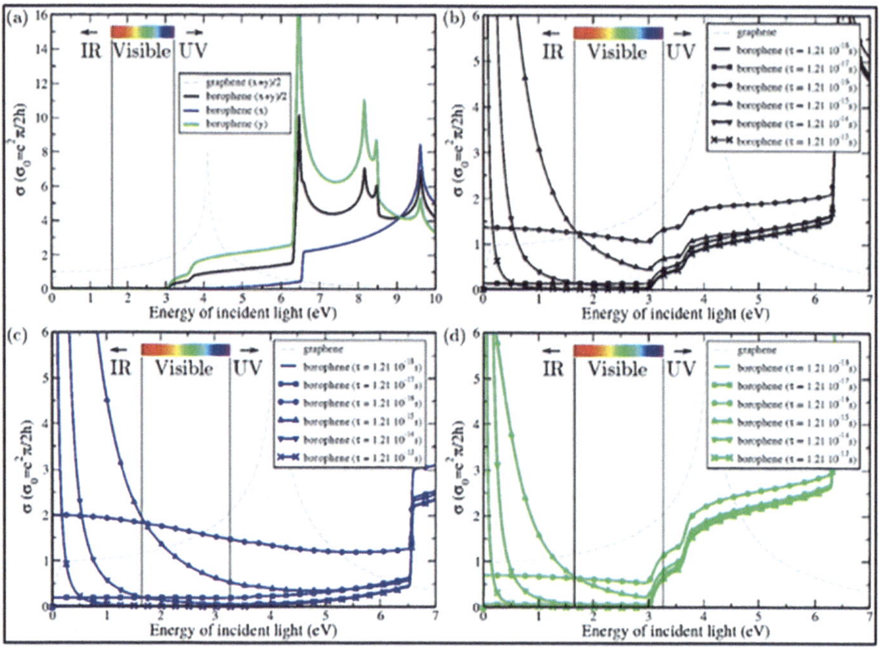

Fig. 3 Optical characteristics of borophene and graphene in different electric field directions are compared and reprinted from [39] copyright (2023), with permission from Elsevier

coefficient. Additionally, the strong light-matter interactions that borophene demonstrates make it advantageous for use in spectroscopy and sensing [35]. Hexagonal BN is transparent in the visible region of the electromagnetic spectrum with a band gap of around 5.5–6.0 eV. It is optically transparent and can be used as a substrate in optical devices or as an insulating layer in optoelectronic applications. Optical properties such as refractive index, extension coefficient, and thickness of the BC can also be determined with the help of a normal reflection spectrum using the Inverse synthesis method. This method involves the utilization of the right dispersion model to fit the measured spectrum to a theoretically generated spectrum. A detailed description of the inverse synthesis method can be seen in the following literature [36–38]. Lherbier et al. reported the real part of optical conductivity ($\sigma_1(E) = E\ \varepsilon_0\varepsilon_2\ (E)/h$, where ε_0 is the vacuum dielectric constant) of borophene is first investigated and is displayed in Fig. 3.

3.3 *Electrical Properties*

The excellent electronic conductivity of boron-based 2D nanosheets makes them suitable for electronic applications. Because borophene has a highly anisotropic electronic band structure, it has highly directional electrical conductivity. Due to its anisotropy, borophene can be used to create novel electronic devices. These materials have high electron mobility, which is a measure of how easily electrons can move through them. In addition, borophene possesses a tunable bandgap, which enables it to be tailored to specific applications. In contrast, h-BN has a very wide bandgap, making it a less electrically conductive or insulating material with a bandgap of nearly 5.9 eV. This implies that h-BN requires very high energy to excite e- to the conduction band from the valence band. At room temperature, it is unable to conduct electricity and has a high electrical resistance. The electrical resistivity of the BN nanosheets decreases rapidly when their thickness decreases while observing through the longitudinal direction. This feature makes it helpful in applications requiring electrical insulation, such as insulating layer coating in electrical devices.

3.4 Magnetic Properties

It has been demonstrated that boron-based 2D nanosheets exhibit interesting magnetic properties due to the presence of boron vacancies within the lattice which leads to unpaired electrons and these unpaired electrons lead to magnetic moment. As a result of these magnetic properties shown by the borophene, make them a potential candidate for use in spintronics. On the other hand, h-BN is a non-magnetic semiconductor material. Some studies suggest that the invasion of defects and doping in the h-BN matrix or system leads to spontaneous magnetization [35]. According to Wu et al. If boron atoms and nitrogen atoms are replaced by carbon atoms, they result in the generation of spontaneous magnetism in 2D h-BN [40]. Vacancy defects in the B or N atoms of h-BN can lead to spontaneous spin magnetization, as demonstrated by the work of Si et al. [41]. A variety of states of BN nanoribbons are shown to exhibit excellent half-metallic magnetism in these studies [42]. In contrast to the above two materials, BC is considered non-magnetic in bulk form. However, some studies suggest that while reducing BC to nanoscale there is a possibility of inducing magnetic properties through various means like structural modification, doping, or defect engineering.

4 Application of Boron-Based 2D Nanosheets for Energy Materials

4.1 Batteries

Batteries convert chemical energy into electrical energy and store it in the form of charge. It is mainly composed of an electrochemical cell having two electrodes that are separated and dipped in an electrolyte. While charging, a voltage is applied across the battery, causing ions to transfer from the positively charged electrode to the negatively charged electrode through the solution [43]. However, during discharge ion diffusion gets reversed and moves from the anode to the cathode leading to a decrease in the voltage until the time it is completely depleted, at which point it must be recharged.

A recent study investigated the use of boron-based 2D nanosheets such as borophene, BC, and BN as an electrode material and catalysts to improve the performance and safety of batteries.

An electrode is a battery component that can act as the source or sink of electric current. A battery normally has two electrodes positive (cathode) and one negative (anode)-each of which is submerged in an electrolyte solution. To enhance the performance of batteries, electrode material should possess a very high surface-to-volume ratio [44], high electron transport [45], reversible capacity [46], and rapid ion diffusion [47]. Boron-based 2D nanosheets possess high metallic character, high anisotropy, great flexibility, and good conduction of heat and electricity [14] which makes it a potential material for the fabrication of secondary batteries more specifically sodium-oxygen batteries, lithium (Li)-ion batteries, and sodium (Na)-ion batteries [14]. Due to its rapid cycle time and performance, Li-ion batteries (LIB) are increasingly popular and widely employed in electric cars, portable gadgets, and grid-scale energy storage. The performance of LIBs, on the other hand, is frequently limited by the electrode materials, which require high energy density, good stability, and fast kinetics [48] (Fig. 4).

Borophene, a two-dimensional material made up of boron atoms, has received a lot of attention in recent years because of its exceptional electronic, thermal, and mechanical properties [49]. Borophene has proved to be a potential material to be used as electrode material in LIBs. Because it is environmentally friendly, has a long cycle life, and has a high energy density [50].

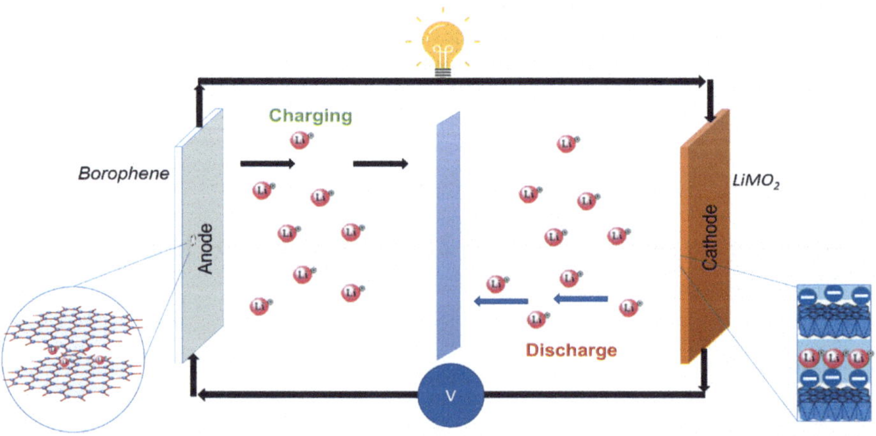

Fig. 4 Borophene-based electrode in Li-ion battery

Also, borophene has several properties that make it a material of interest for LIB electrodes. It has a theoretical specific capacity of up to 1720 mAh/g [51], which is significantly greater than traditional electrode materials such as graphite 372 mAh/g due to the unique electronic structure of borophene, which enables reversible intercalation of lithium ions. Due to its high electrical conductivity, borophene allows for rapid electron transfer during the charge and discharge processes, is mechanically flexible, and can withstand repeated expansion and contraction cycles during battery operation [52].

Recent research has investigated the use of borophene as electrode material in LIBs. One study, for example, used DFT calculations to predict the electrochemical properties of borophene as an anodic material in LIBs [53]. According to the findings, Borophene has a high Li storage capacity, a low Li diffusion barrier, and good electronic conductivity, indicating its potential as a high-performance anode material. Another study synthesized borophene nanosheets experimentally and demonstrated their use as a high-capacity anode material in LIBs. After 200 cycles, the borophene 2D nanosheets had a reversible specific capacity of 634 mAh-g^{-1}, which is significantly higher than graphite anodes [54].

BN is a synthetic inorganic compound composed of boron and nitrogen arranged in a 2D nanosheet structure. It can be synthesized via various methods, including chemical vapor deposition, ball milling, and sol–gel methods [55]. Due to its 2D structure, it exhibits great mechanical strength, high thermal stability, and thermal conductivity. BN can be made chemically inert, making it a potential material to be used in harsh environments [56]. In recent years, researchers came up with using BN and its composites as electrode materials for batteries due to their, long cycle life, low environmental impact, and high energy density making it a preferential material for applications such as the electronic and automotive sectors [57].

The limitations with the use of traditional graphite-based electrodes in LIBs including their low energy density, low chemical resistance, and low thermal stability can be resolved by using BN and based composites [58]. The excellent electrochemical performance, high thermal stability, and improved safety characteristics make BN a superior material to graphite for electrode material. In addition, in recent years BN-based composites have been developed to improve the electrochemical performance of BN electrodes [59], and according to the outcomes BN-carbon hybrid composites have been shown to have superior electrochemical performance compared to pure BN electrodes with better cycle stability and improved specific capacity. Furthermore, other type of BN-based composites also proves themselves a promising material such as BN/metal oxide and BN/graphene composites which improve the overall electrochemical performance of BN-based materials [60].

B_4C is known for its hardness, high melting point, and good chemical stability. It has been used in various applications, including armor materials, nuclear industry, and electronics. Due to its high specific capacity, good cycling stability, and low cost, its use as electrode material in batteries has frequently increased [61].

BC-based electrodes exhibit a high specific capacity of up to 2000 mAh/g which is much higher than most commercial electrode materials [62]. However, its lower electrical conductivity and poor rate capability limit its usage in commercial batteries.

To resolve such shortcomings, BC can be combined with more electrically conductive additives, carbon nanotubes (CNTs), graphene, conducting polymers, and metal nanoparticles, to form composites [63].

B_4C composites show increased electrochemical performance. For example, in a study B_4C-graphene composite has been reported high specific capacity of 414 $mAhg^{-1}$ with a good cycle stability of up to 100 cycles. In Another study, researchers reported that a B_4C-CNT composite showed a high specific capacity of 542 $mAhg^{-1}$ and good rate capability. The use of B4Ccomposites as electrode materials also offers other advantages. For instance, it can act as a host material for other active materials, such as lithium sulfide or silicon, to improve their cycle stability and rate capability [64]. Additionally, B_4C composites can be fabricated using low-cost and scalable methods, such as ball milling or spray drying, making them attractive for commercial production.

Other Boron-Derived Nanosheets

The application of boron phosphide (BP) 2D nanosheets as material for LIB has been investigated due to their high theoretical capacity and excellent cycle stability. Its ability to store lithium ions by forming a solid-electrolyte interphase (SEI) on its surface prevents further reaction with the electrolyte and improves its cycle stability. Moreover, BP nanosheets have good electron conductivity, which helps in better charge transfer and can enhance their rate capability [65].

Boron oxide (B_2O_3) nanosheets can be used as cathodic material for Na-ion batteries because of their high theoretical capacity and low cost. These nanosheets can store Na ions by forming a reversible intercalation compound, which can provide a high specific capacity and high surface area and good electronic conductivity of B_2O_3 nanosheets can improve the electrochemical performance [66].

Boron carbonitrides (BCN) 2D nanosheets have been investigated as cathodes for lithium-sulfur batteries due to their good theoretical capacity and good chemical stability. BCN nanosheets can trap polysulfides and prevent their dissolution into the electrolyte, which can improve the cycle stability of the batteries. Additionally, BCN nanosheets possess good electrical conductivity attributed to their good rate capability [67].

4.2 Catalyst

In Borophene, boron atoms are organized in a hexagonal lattice form, having promising features for a variety of materials science applications. The potential of borophene as a catalyst for energy storage devices, particularly batteries, has piqued the interest of academics in recent years.

The incorporation of borophene as a catalyst in batteries may improve their energy storage capacity, cycle stability, and rate capability greatly. Due to its large surface area and high electrical conductivity allows it efficient electron transport and increases the catalytic activity of electrode materials in batteries. Furthermore, it has unique electronic features, such as high reactivity and variable work function which makes it a suitable candidate for many electrochemical processes involved in battery operation. For example, in rechargeable metal-air batteries such as Li-O_2 and Zn-air batteries, borophene may operate as an effective catalyst for the O_2 reduction reaction (ORR) and O_2 evolution reaction (OER). Recent research has demonstrated that the use of borophene as a catalyst in electrode materials may increase overall battery performance. In recent work, researchers created a hybrid electrode material by depositing borophene on a graphene oxide (GO) nanosheet. With a specific capacity of 1469.35 $mAhg^{-1}$ and remarkable cycle stability, the resultant material demonstrated improved electrochemical performance [68].

In another work, borophene was used as a catalyst in a lithium-sulfur (Li–S) battery, which enhanced battery performance by improving reaction kinetics and decreasing polarisation during charge–discharge cycles. The borophene-based catalyst was also more stable since it inhibited sulfur intermediate dissolution and the generation of insulating lithium sulfides on the electrode surface [69].

The use of BN as a catalyst not only enhances the electrochemical performance of the electrodes but also improves the stability and durability of the electrode by preventing the formation of unwanted by-products and reducing the degradation of the electrode material. It can also act as a mediator to enhance the redox reactions and increase the electron transfer rate at the electrode/electrolyte interface [70]. One potential application of BN as a catalyst in batteries is in lithium-sulfur (Li–S) batteries. Li–S batteries have a high theoretical energy density but suffer from poor cycle life and low-rate capability. BN can be used as a catalyst to improve the electrochemical performance of the sulfur cathode by facilitating the conversion of sulfur to lithium sulfide, which is a more stable and conductive intermediate. BN can also be capable of preventing the formation of polysulfides, leading to the degradation of the sulfur cathode. Another potential application of BN as a catalyst is in lithium-oxygen (Li-O_2) batteries [71]. These batteries have a high theoretical energy density but due to the formation of unwanted products, they result in low efficiency and poor cycle life. BN-based catalyst can be used to enhance the efficiency of the ORR and OER at the cathode, preventing the formation of unwanted by-products such as lithium peroxide and lithium oxide and increasing the shelf life of the electrode.

Various features of B_4C that make it an appealing option for use as a catalyst in batteries are Great hardness, high thermal, and chemical stability, and large surface area. B_4C can be used as a catalyst in batteries to improve the electrochemical performance of the electrodes. By lowering the activation energy necessary for the processes and improving the kinetics of the electrochemical reactions. It may also increase the electrode's stability and endurance by limiting the development of undesirable by-products and lowering electrode material deterioration. It may also operate as a mediator to improve redox processes and speed up electron transport at the electrode/electrolyte contact. Li-ion batteries are one possible use for B_4C as a catalyst that improves the electrochemical processes at the electrode/electrolyte interface, the efficiency of the charge/discharge process, and the electrode materials' durability [61].

Other Boron Derived Nanosheets

The boron phosphide (BP) nanosheet catalyst can be used in OER in water-splitting electrochemical cells. The OER is a critical reaction in renewable energy conversion technologies, such as rechargeable metal-air batteries and hydrogen production. BP-based catalysts can enhance the OER performance due to their high electrical conductivity, high surface area, and unique crystal structure attributed to the improved durability and stability of the catalysts during the OER process [72].

The Application of a Boron oxide (B_2O_3) catalyst has been studied for the ORR in metal-air batteries. ORR converts chemical energy to electrical energy by reacting the metal with oxygen from the air. Due to their high surface area, high conductivity, and unique crystal structure, B2O3 nanosheets can improve the ORR performance. Additionally, B_2O_3 nanosheets can also improve the durability and stability of the catalysts during the ORR process [73].

4.3 Supercapacitor

Supercapacitors or ultracapacitors are energy storage devices that can quickly charge and discharge large amounts of electrical energy. They have numerous applications such as transportation, renewable energy, and electronics for energy storage. The working principle and building blocks of a general supercapacitor are shown in Fig. 5.

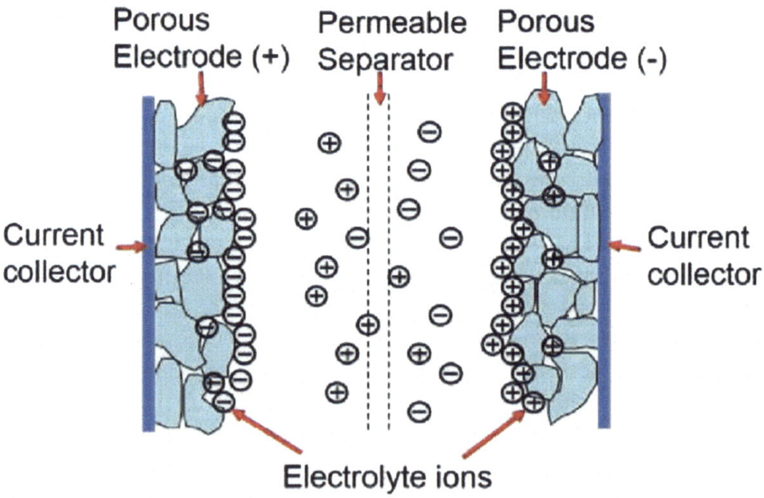

Fig. 5 The schematic illusion of working and building blocks of supercapacitor reprints from [74] copyright (2023), with permission from Elsevier

Supercapacitor electrodes based on Boron-based materials, such as borophene, BN, and BC, proved themselves a potential material due to their high electrical conductivity, high surface area, and faster charge–discharge cycle [62]. These materials have unique 2D structures that allow for efficient charge transfer and storage, resulting in high capacitance values. Additionally, these materials are lightweight, mechanically robust, and have good chemical stability, making them suitable for use in various supercapacitor applications [75].

Due hexagonal 2D structure of borophene, makes it a promising candidate for use in supercapacitor applications. One of the most important properties of borophene is its high electrical conductivity which allows it to efficiently conduct and store electrical charge, which is crucial for the efficient operation of supercapacitors [76].

Borophene has a high surface area which enables it to have more surface sites for charge storage and improves its capacitance. In addition to its electrical properties, borophene has good mechanical properties and high flexibility that facilitate its formability into desired shapes making it a versatile material for flexible energy storage devices [77]. Also, borophene has high durability, so it can withstand the repeated charge–discharge cycles that occur during the operation of supercapacitors, making it an ideal material for long-lasting energy storage devices [78]. Its electronic and mechanical properties can be tuned by forming defects, such as vacancies and adatom. Furthermore, borophene has a low density attributed to providing high energy storage capacity without adding too much weight to the device which is particularly important for applications such as electric vehicles, where weight is a critical factor in determining the overall efficiency and performance of the system [3].

BN has a very large surface area, which is crucial for better supercapacitor performance. The large surface area offers more locations for electrolyte ion adsorption and desorption, resulting in better capacitance and energy storage [79]. It is extremely conductive, due to which it is useful for efficient charge transfer and energy storage and its exceptional thermal stability and chemical inertness make it an ideal material for supercapacitor electrodes since it can survive extreme working conditions without deteriorating or decomposing. Several approaches for manufacturing BN-based materials for supercapacitor applications have been investigated by researchers. One method is to use a simple hydrothermal procedure to create BN nanotubes with high specific capacitance and great cycle stability [80]. To improve supercapacitor performance, further options include employing BN/graphene composites or BN/metal oxide hybrids. When compared to standard carbon-based electrodes, these materials may provide higher energy density, power density, and stability [81].

BC is an excellent material for supercapacitor electrodes. It has a large surface area with good conductivity, which offers better capacitance and energy storage and efficient charge transfer & energy storage. B_4C in addition to its large surface area and conductivity, is chemically inert, stable at high temperatures, and has low density. Several approaches for manufacturing B_4C-based materials for supercapacitor applications have been investigated by researchers. One way includes preparing B_4C nanoparticles using a simple sol–gel procedure, which may result in high specific capacitance and great cycle stability. Another method is to use B_4C/

graphene composites to improve supercapacitor performance [64]. The use of B_4C in supercapacitors is still in its early phase of development. However, the unusual mix of characteristics of B_4C makes it an intriguing contender for use in high-power applications.

Other Boron Derived Nanosheets

The properties of Boron phosphide (BP) nanosheets have been investigated as electrode materials for supercapacitors. Due to its high electrical conductivity, high specific capacitance, rate capability, and excellent cycling stability BP nanosheets can store charges by electrostatic adsorption [82].

Boron oxide (B_2O_3) nanosheets have been studied as electrode materials for supercapacitors due to their high surface area, high electrical conductivity, and good electrochemical stability. B_2O_3 nanosheets can store charges by electrostatic adsorption, which can provide a high specific capacitance. In addition, B_2O_3 nanosheets have a high surface area and good electronic conductivity, which can improve their electrochemical performance [79].

Boron carbonitride (BCN) nanosheets have been investigated as electrode materials for supercapacitors due to their high specific capacitance, excellent rate capability, and good chemical stability. BCN nanosheets can store charges by a combination of electrostatic adsorption and pseudo-capacitance, which can provide a high specific capacitance. In addition, BCN nanosheets have good electron conductivity, which can improve their rate capability [83].

4.4 Hydrogen Storage/Fuel Cell

It has been seen that hydrogen has high energy density and shows environmental friendliness. However, the major obstacle in this sector to use hydrogen as fuel is its storage for practical applications. In the past few years, boron-based 2D materials such as borophene, BN (BN), and B_4C have gained significant attention due to their significant properties such as high hydrogen binding energy, good thermal stability, high electrical conductivity, and facile synthesis. The application of boron-based 2D materials as hydrogen storage materials has thus been the subject of in-depth research. To gain insight into the hydrogen storage capability of these boron-based 2D materials, researchers have used a variety of methods including theoretical computational modeling and experimental studies. These investigations have produced positive findings with some materials showing high hydrogen storage capacities, substantial stability, and rapid hydrogen uptake and release rates.

Borophene is a highly anisotropic 2D allotrope of boron that may be formed into a variety of shapes, including hexagonal, nanoribbons, and triangular. A potential material for hydrogen storage, borophene has been proven to have great stability and strong mechanical characteristics. Recent research has shown that borophene is capable of adsorbing hydrogen through both physisorption and chemisorption pathways, with the latter mechanism being responsible for the majority of hydrogen storage. Borophene is a good contender for use in practical applications because it has demonstrated hydrogen storage capabilities of up to 15.6% by weight at room temperature and atmospheric pressure. Suleyman et al. compared the hydrogen adsorption capacities of α-boron sheets with graphene to obtain some encouraging results [84]. Hydrogen binding energies were also investigated by doping boron sheets with alkali metals (AM). H_2 binding energies and densities of Li, Na, and K-doped boron sheets have been computed. Maximum hydrogen adsorption storage capacity was determined to be 10.75 wt% for Li-doped systems, with 0.15 eV binding energy per H_2 molecule, whereas it was 8.36 wt% for Na and 2.39 wt% for K. Graphene doped with AM atoms was found to be unstable in comparison to the bulk metal, leading the researchers to conclude that boron nanostructures doped with AM were more effective than their carbon counterparts. Li et al. used density functional theory (DFT) simulations to investigate the hydrogen storage and electrical characteristics of Li-doped borophene. This study is unique in that it employs three different approaches to calculating adsorption energies, namely the local density approximation (LDA) functional using the Perdew–Wang method, the generalized gradient approximation (GGA) functional using the Perdew-Burke-Ernzerhof (PBE) functional, and the DFT-D2 method of Grimme, which takes into account van der Waals (vdW) interactions. As DFT calculations only provide for the structural and electrical characterizations of borophene, we needed to use three different approaches to determine the long-range weak van der Waal interactions between ions and electrons [85].

BN is a hexagonal sheet-like material consisting of alternative nitrogen and boron atoms and has a unique honeycomb-like structure just like graphene. BN is highly porous and has a high surface area which makes it an alternative material for hydrogen storage. Studies show that just like borophene, BN can store hydrogen through both chemisorption and physisorption mechanisms. At room temperature and atmospheric pressure, BN shows 7–12% by weight hydrogen storage capacity, which represents strong binding between hydrogen and boron (Table 2).

5 Conclusion

The potential of boron-based materials for batteries, catalysis, and fuel cells holds tremendous promise for advancing sustainable energy technologies. Throughout the chapter, we see how the structure and unique properties of boron-based compounds show potential for these applications. BBNS offers high capacity, excellent cycling stability, and enhanced charge/discharge kinetics, enabling the development of efficient and long-lasting batteries and supercapacitors. These materials show great potential for overcoming the limitations of traditional energy storage technologies and enabling the widespread adoption of renewable energy sources. Also, because of their variable chemical and electrical characteristics, they are flexible catalysts capable of increasing reaction rate, selectivity, and overall energy efficiency. Furthermore, boron-based catalysts have great stability and low toxicity, making them more ecologically benign than traditional 2D catalysts. However, even with the

Table 2 Boron-based materials and their potential energy applications

S.N	Material	Potential application	Specific capacity (experimental/theoretical)	References
1	Borophene	Libs	3306 mA hg^{-1}	[86]
2	Borophene oxide	LIBs	2137 mA hg^{-1}	[87]
3	Borophene oxide	Sodium-ion batteries	1425 mA hg^{-1}	[87]
4	β12 borophene	LIBs	1984 mA hg^{-1}	[50]
5	χ3 borophene	LIBs, NIBs	1240 mA hg^{-1}	[50]
6	Flat borophene film	Batteries	1640 – 2480 mA hg^{-1}	[88]
7	Li decorated borophene	Hydrogen storage	13.7 wt%	[89]
8	Charged borophene	Hydrogen storage	6.5 wt%	[90]
9	Ca decorated boron sheet	Hydrogen storage	8.92 wt%	[91]
10	Nitrogen-decorated boron sheet	Hydrogen storage	6.22 wt%	[92]

great advances made in boron-based materials, further research, and development are required to fully realize their promise. To promote large-scale commercialization, issues such as synthesis scalability, cost-effectiveness, and long-term stability must be addressed. Furthermore, efforts should be made to understand the fundamental mechanisms behind their distinctive features to lead the creation of innovative boron-based materials with specific functionality.

References

1. Huang, Z., Wang, S., Dewhurst, R.D., Ignat'ev, N.V., Finze, M., Braunschweig, H.: Boron: its role in energy-related processes and applications. Angew. Chem. Int. Ed.. Wiley-VCH Verlag (2020). https://doi.org/10.1002/anie.201911108
2. Xie, Z., Meng, X., Li, X., Liang, W., Huang, W., Chen, K., Zhang, H.: Two-dimensional borophene: properties, fabrication, and promising applications. Research (2020). https://doi.org/10.34133/2020/2624617
3. Hou, C., Tai, G., Liu, Y., Wu, Z., Liang, X., Liu, X.: Borophene-based materials for energy, sensors and information storage applications. Nano Res. Energy (2023). https://doi.org/10.26599/nre.2023.9120051
4. Preobrajenski, A.B., Lyalin, A., Taketsugu, T., Vinogradov, N.A., Vinogradov, A.S.: Honeycomb boron on Al(111): from the concept of borophene to the two-dimensional boride. ACS Nano **15**(9), 15153–15165 (2021). https://doi.org/10.1021/acsnano.1c05603
5. Cao, W., Jiang, J., Xie, X., Pal, A., Chu, J.H., Kang, J., Banerjee, K.: 2-D layered materials for next-generation electronics: opportunities and challenges. IEEE Trans. Electron Devices **65**(10), 4109–4121 (2018). https://doi.org/10.1109/TED.2018.2867441
6. Gupta, S., Kutana, A., Yakobson, B.I.: Dirac cones and nodal line in borophene. J. Phys. Chem. Lett. **9**(11), 2757–2762 (2018). https://doi.org/10.1021/acs.jpclett.8b00640

7. Ranjan, P., Lee, J.M., Kumar, P., Vinu, A.: Borophene: new sensation in flatland. Adv. Mater. **32**(34) (2020). https://doi.org/10.1002/adma.202000531
8. Cheng, T., Lang, H., Li, Z., Liu, Z., Liu, Z.: Anisotropic carrier mobility in two-dimensional materials with tilted Dirac cones: theory and application. Phys. Chem. Chem. Phys. **19**(35), 23942–23950 (2017). https://doi.org/10.1039/c7cp03736h
9. Vatanpour, V., Naziri Mehrabani, S.A., Keskin, B., Arabi, N., Zeytuncu, B., Koyuncu, I.: A comprehensive review on the applications of boron nitride nanomaterials in membrane fabrication and modification. Ind. Eng. Chem. Res. Am. Chem. Soc. (2021). https://doi.org/10.1021/acs.iecr.1c02102
10. Muraleedharan, M.G., Kent, P.R.C.: Novel boron nitride MXenes as promising energy storage materials. Nanoscale **14**(25), 9086–9096 (2022). https://doi.org/10.1039/d2nr02807g
11. Wang, J., Zhang, L., Wang, L., Lei, W., Wu, Z.S.: Two-dimensional boron nitride for electronics and energy applications. In: Energy and Environmental Materials. John Wiley and Sons Inc. (2022). https://doi.org/10.1002/eem2.12159
12. Wu, Y., Cao, Y., Wu, Y., Li, D.: Neutron shielding performance of 3D-Printed boron carbide PEEK composites. Materials **13**(10). https://doi.org/10.3390/ma13102314
13. Fan, D., Lu, S., Guo, Y., Hu, X.: Two-dimensional stoichiometric boron carbides with unexpected chemical bonding and promising electronic properties. J. Mater. Chem. C **6**(7), 1651–1658 (2018). https://doi.org/10.1039/c7tc04505k
14. Rubab, A., Baig, N., Sher, M., Sohail, M.: Advances in ultrathin borophene materials. Chem. Eng. J.. Elsevier B.V (2020). https://doi.org/10.1016/j.cej.2020.126109
15. Dean, C.R., Young, A.F., Meric, I., Lee, C., Wang, L., Sorgenfrei, S., Hone, J.: Boron nitride substrates for high-quality graphene electronics. Nat. Nanotechnol. **5**(10), 722–726 (2010). https://doi.org/10.1038/nnano.2010.172
16. Mannix, A.J., Zhou, X.F., Kiraly, B., Wood, J.D., Alducin, D., Myers, B.D., Guisinger, N.P.: Synthesis of borophenes: anisotropic, two-dimensional boron polymorphs. Science **350**(6267), 1513–1516 (2015). https://doi.org/10.1126/science.aad1080
17. Yu, C., Zhang, J., Tian, W., Fan, X., Yao, Y.: Polymer composites based on hexagonal boron nitride and their application in thermally conductive composites. Royal Society of Chemistry, RSC Advances (2018). https://doi.org/10.1039/c8ra02685h
18. Gautam, C., Chelliah, S.: Methods of hexagonal boron nitride exfoliation and its functionalization: covalent and non-covalent approaches. In: RSC Advances. Royal Society of Chemistry (2021). https://doi.org/10.1039/d1ra05727h
19. Horiba, K., Ohguchi, H., Kumigashira, H., Oshima, M., Ono, K., Nakagawa, N., Koinuma, H.: A high-resolution synchrotron-radiation angle-resolved photoemission spectrometer with in situ oxide thin film growth capability. Rev. Sci. Instrum. **74**(7), 3406–3412 (2003). https://doi.org/10.1063/1.1584093
20. Marcelo, J., Lopes, J.: MBE growth of graphene (2019)
21. Mazaheri, A., Javadi, M., Abdi, Y.: Chemical vapor deposition of two-dimensional boron sheets by thermal decomposition of diborane. ACS Appl. Mater. Interfaces. **13**(7), 8844–8850 (2021). https://doi.org/10.1021/acsami.0c22580
22. Kiraly, B., Liu, X., Wang, L., Zhang, Z., Mannix, A.J., Fisher, B.L., Guisinger, N.P.: Borophene synthesis on Au(111). ACS Nano **13**(4), 3816–3822 (2019). https://doi.org/10.1021/acsnano.8b09339
23. Zhong, Q., Kong, L., Gou, J., Li, W., Sheng, S., Yang, S., Chen, L.: Synthesis of borophene nanoribbons on Ag(110) surface. Phys. Rev. Mater. **1**(2) (2017). https://doi.org/10.1103/PhysRevMaterials.1.021001
24. Wu, Z., Tai, G., Liu, R., Shao, W., Hou, C., Liang, X.: Synthesis of borophene on quartz towards hydroelectric generators. J. Mater. Chem. A **10**(15), 8218–8226 (2022). https://doi.org/10.1039/d1ta10855g
25. Sutter, P., Sutter, E.: Large-scale layer-by-layer synthesis of borophene on Ru(0001). Chem. Mater. **33**(22), 8838–8843 (2021). https://doi.org/10.1021/acs.chemmater.1c03061
26. Lei, W., Mochalin, V.N., Liu, D., Qin, S., Gogotsi, Y., Chen, Y.: Boron nitride colloidal solutions, ultralight aerogels and freestanding membranes through one-step exfoliation and functionalization. Nat. Commun. **6** (2015). https://doi.org/10.1038/ncomms9849

27. Coleman, J.N., Lotya, M., O'neill, A., Bergin, S.D., King, P.J., Khan, U., Nicolosi, V.: Two-Dimensional Nanosheets Produced by Liquid Exfoliation of Layered Materials (n.d.). http://science.sciencemag.org/
28. Wang, X., Hooper, T.N., Kumar, A., Priest, I.K., Sheng, Y., Samuels, T.O.M., Warner, J.H.: Oligomeric aminoborane precursors for the chemical vapour deposition growth of few-layer hexagonal boron nitride. CrystEngComm **19**(2), 285–294 (2017). https://doi.org/10.1039/c6ce02006b
29. Oh, H., Jo, J., Tchoe, Y., Yoon, H., Lee, H.H., Kim, S.S., Yi, G.C.: Centimeter-sized epitaxial h-BN films. NPG Asia Mater. **8**(11), 1–8 (2016). https://doi.org/10.1038/am.2016.178
30. Liu, H., You, C.Y., Li, J., Galligan, P.R., You, J., Liu, Z., Luo, Z.: Synthesis of hexagonal boron nitrides by chemical vapor deposition and their use as single photon emitters. Nano Mater. Sci. **3**(3), 291–312 (2021). https://doi.org/10.1016/j.nanoms.2021.03.002
31. Yadav, S., Sadique, M.A., Kaushik, A., Ranjan, P., Khan, R., Srivastava, A.K.: Borophene as an emerging 2D flatland for biomedical applications: current challenges and future prospects. J. Mater. Chem. B. R. Soc. Chem. (2022). https://doi.org/10.1039/d1tb02277f
32. Wang, Z.Q., Lü, T.Y., Wang, H.Q., Feng, Y.P., Zheng, J.C.: Review of borophene and its potential applications. In: Frontiers of Physics. Higher Education Press (2019). https://doi.org/10.1007/s11467-019-0884-5
33. Domnich, V., Reynaud, S., Haber, R.A., Chhowalla, M.: Boron carbide: structure, properties, and stability under stress. J. Am. Ceram. Soc. **94**(11), 3605–3628 (2011). https://doi.org/10.1111/j.1551-2916.2011.04865.x
34. dos Santos, O.F.P., Azevedo, S.: Effect of strain on the mechanical and electronic properties of h-BN nanoribbons with 558 linear defect. Phys. E: Low-Dimens. Syst. Nanostructures **116** (2020). https://doi.org/10.1016/j.physe.2019.113771
35. Wang, J., Ma, F., Sun, M.: Graphene, hexagonal boron nitride, and their heterostructures: properties and applications. Royal Society of Chemistry, RSC Advances (2017). https://doi.org/10.1039/c7ra00260b
36. Minkov, D.A.: Calculation of the optical constants of a thin layer upon a transparent substrate from the reflection spectrum Printed in the UK. J. Phys. D: Appl. Phys. **22** (1989). http://iopscience.iop.org/0022-3727/22/8/021
37. Jena, S., Tokas, R.B., Tripathi, S., Rao, K.D., Udupa, D.V., Thakur, S., Sahoo, N.K.: Influence of oxygen partial pressure on microstructure, optical properties, residual stress and laser induced damage threshold of amorphous HfO2 thin films. J. Alloy. Compd. **771**, 373–381 (2019). https://doi.org/10.1016/j.jallcom.2018.08.327
38. Jena, S., Tokas, R.B., Thakur, S., Sahoo, N.K.: Optical properties of electron beam evaporated ZrO2:10 %SiO2 thin films: dependence on structure. Indian J. Phys. **90**(8), 951–957 (2016). https://doi.org/10.1007/s12648-016-0831-z
39. Lherbier, A., Rafael Botello-Méndez, A., Charlier, J.C.: Electronic and optical properties of pristine and oxidized borophene. 2D Mater. **3**(4) (2016). https://doi.org/10.1088/2053-1583/3/4/045006
40. Wu, R.Q., Liu, L., Peng, G.W., Feng, Y.P.: Magnetism in BN nanotubes induced by carbon doping. Appl. Phys. Lett. **86**(12), 1–3 (2005). https://doi.org/10.1063/1.1890477
41. Si, M.S., Xue, D.S.: Magnetic properties of vacancies in a graphitic boron nitride sheet by first-principles pseudopotential calculations. Phys. Rev. B-Condens. Matter Mater. Phys. **75**(19) (2007). https://doi.org/10.1103/PhysRevB.75.193409
42. Lin, L., Jing, L., Lu, W., Guangfu, L., Jing, Z., Rui, Q., Wai, N.M.: Magnetic properties of fully bare and half-bare boron nitride nanoribbons. J. Phys. Chem. C **113**(6), 2273–2276 (2009). https://doi.org/10.1021/jp8079827
43. Winter, M., Brodd, R.J.: What are batteries, fuel cells, and supercapacitors? Chem. Rev. **104**(10), 4245–4269 (2004). https://doi.org/10.1021/cr020730k
44. Wang, K.X., Li, X.H., Chen, J.S.: Surface and interface engineering of electrode materials for lithium-ion batteries. Adv. Mater. **27**(3), 527–545 (2015). https://doi.org/10.1002/adma.201402962

45. Lee, S., Hong, J., Jung, S.K., Ku, K., Kwon, G., Seong, W.M., Kang, K.: Charge-transfer complexes for high-power organic rechargeable batteries. Energy Storage Mater. **20**, 462–469 (2019). https://doi.org/10.1016/j.ensm.2019.05.001
46. Chadha, U., Bhardwaj, P., Padmanaban, S., Suneel, R.M., Milton, K., Subair, N., Kumar, A.: Review—contemporary progresses in carbon-based electrode material in Li-S batteries. J. Electrochem. Soc. **169**(2), 020530 (2022). https://doi.org/10.1149/1945-7111/ac4cd7
47. Chen, J., Cheng, F.: Combination of lightweight elements and nanostructured materials for batteries. Acc. Chem. Res. **42**(6), 713–723 (2009). https://doi.org/10.1021/ar800229g
48. Lu, Y., Chen, J.: Prospects of organic electrode materials for practical lithium batteries. Nat. Rev. Chemistry. Nat. Res. (2020). https://doi.org/10.1038/s41570-020-0160-9
49. Yang, R., Sun, M.: Borophenes: monolayer, bilayer and heterostructures. J. Mater. Chem. C. R. Soc. Chem. (2023). https://doi.org/10.1039/d3tc00974b
50. Zhang, X., Hu, J., Cheng, Y., Yang, H.Y., Yao, Y., Yang, S.A.: Borophene as an extremely high capacity electrode material for Li-ion and Na-ion batteries. Nanoscale **8**(33), 15340–15347 (2016). https://doi.org/10.1039/c6nr04186h
51. Jiang, H.R., Lu, Z., Wu, M.C., Ciucci, F., Zhao, T.S.: Borophene: a promising anode material offering high specific capacity and high rate capability for lithium-ion batteries. Nano Energy **23**, 97–104 (2016). https://doi.org/10.1016/j.nanoen.2016.03.013
52. Bree, G., Hao, H., Stoeva, Z., John Low, C.T.: Monitoring state of charge and volume expansion in lithium-ion batteries: an approach using surface mounted thin-film graphene sensors. RSC Adv. **13**(10), 7045–7054 (2023). https://doi.org/10.1039/d2ra07572e
53. Liu, J., Zhang, C., Xu, L., Ju, S.: Borophene as a promising anode material for sodium-ion batteries with high capacity and high rate capability using DFT. RSC Adv. **8**(32), 17773–17785 (2018). https://doi.org/10.1039/c8ra01942h
54. Makaremi, M., Mortazavi, B., Singh, C.V.: 2D Hydrogenated Graphene-like Borophene as a High Capacity Anode Material for Improved Li/Na Ion Batteries: A First Principles Study (n.d.)
55. Ahmad, P., Khandaker, M.U., Mohd Amin, Y., Raza Khan, Z.: Influence of growth duration on size and morphology of boron nitride nanotubes grown via chemical vapor deposition technique. J. Phys. Chem. Solids **85**, 226–232 (2015). https://doi.org/10.1016/j.jpcs.2015.06.003
56. Shen, H., Guo, J., Wang, H., Zhao, N., Xu, J.: Bioinspired modification of h-BN for high thermal conductive composite films with aligned structure. ACS Appl. Mater. Interfaces **7**(10), 5701–5708 (2015). https://doi.org/10.1021/am507416y
57. Asenbauer, J., Eisenmann, T., Kuenzel, M., Kazzazi, A., Chen, Z., Bresser, D.: The success story of graphite as a lithium-ion anode material-fundamentals, remaining challenges, and recent developments including silicon (oxide) composites. Sustain. Energy Fuels. R. Soc. Chem. (2020). https://doi.org/10.1039/d0se00175a
58. An, S.J., Li, J., Daniel, C., Mohanty, D., Nagpure, S., Wood, D.L.: The state of understanding of the lithium-ion-battery graphite solid electrolyte interphase (SEI) and its relationship to formation cycling. In: Carbon. Elsevier Ltd. (2016). https://doi.org/10.1016/j.carbon.2016.04.008
59. Fatima, I., Fayyaz, O., Yusuf, M.M., Al Ashraf, A., Shakoor, R.A.: Enhanced electrochemical and mechanical performance of BN reinforced Ni-P based nanocomposite coatings. Diam. Relat. Mater. **130** (2022). https://doi.org/10.1016/j.diamond.2022.109454
60. Huang, T., Zeng, X., Yao, Y., Sun, R., Meng, F., Xu, J., Wong, C.: Boron nitride@graphene oxide hybrids for epoxy composites with enhanced thermal conductivity. RSC Adv. **6**(42), 35847–35854 (2016). https://doi.org/10.1039/c5ra27315c
61. Song, S., Xu, W., Cao, R., Luo, L., Engelhard, M.H., Bowden, M.E., Zhang, J.G.: B4C as a stable non-carbon-based oxygen electrode material for lithium-oxygen batteries. Nano Energy **33**, 195–204 (2017). https://doi.org/10.1016/j.nanoen.2017.01.042
62. Avcıoğlu, S., Buldu-Akturk, M., Erdem, E., Kaya, F., Kaya, C.: Boron carbide as an electrode material: tailoring particle morphology to control capacitive behaviour. Materials **16**(2) (2023). https://doi.org/10.3390/ma16020861

63. Tan, Y., Luo, H., Zhang, H., Peng, S.: Graphene nanoplatelet reinforced boron carbide composites with high electrical and thermal conductivity (2016)
64. Chang, Y., Sun, X., Ma, M., Mu, C., Li, P., Li, L., Tian, Y.: Application of hard ceramic materials B4C in energy storage: design B4C@C core-shell nanoparticles as electrodes for flexible all-solid-state micro-supercapacitors with ultrahigh cyclability. Nano Energy **75** (2020). https://doi.org/10.1016/j.nanoen.2020.104947
65. Banerjee, S., Periyasamy, G., Pati, S.K.: Possible application of 2D-boron sheets as anode material in lithium ion battery: a DFT and AIMD study. J. Mater. Chem. A **2**(11), 3856–3864 (2014). https://doi.org/10.1039/c3ta14041e
66. Sanglay, G.D.D., Garcia, J.S., Palaganas, M.S., Sorolla, M., See, S., Limjuco, L. A., Ocon, J.D.: Borate-based compounds as mixed polyanion cathode materials for advanced batteries. Molecules (2022). MDPI. https://doi.org/10.3390/molecules27228047
67. Jiang, R., Jiang, M., Huang, Z., Wang, J., Kuang, Y., Fu, C.: Constructing light-weight polar boron-doped carbon nitride nanosheets with increased active sites and conductivity for high performance lithium-sulfur batteries. Int. J. Hydrogen Energy **45**(29), 14940–14952 (2020). https://doi.org/10.1016/j.ijhydene.2020.03.232
68. Yu, J., Zhou, M., Yang, M., Yang, Q., Zhang, Z., Zhang, Y.: High-performance borophene/graphene heterostructure anode of lithium-ion batteries achieved via controlled interlayer spacing. ACS Appl. Energy Mater. **3**(12), 11699–11705 (2020). https://doi.org/10.1021/acsaem.0c01808
69. Rao, D., Liu, X., Yang, H., Zhang, L., Qiao, G., Shen, X., Lu, R.: Interfacial competition between a borophene-based cathode and electrolyte for the multiple-sulfide immobilization of a lithium sulfur battery. J. Mater. Chem. A **7**(12), 7092–7098 (2019). https://doi.org/10.1039/c8ta12176a
70. Li, Y., Guo, S.: Material design and structure optimization for rechargeable lithium-sulfur batteries. In: Matter. Cell Press (2021). https://doi.org/10.1016/j.matt.2021.01.012
71. Chowdhury, C., Datta, A.: Doped boron nitride surfaces: Potential metal free bifunctional catalysts for non-aqueous Li-O2 batteries. Phys. Chem. Chem. Phys. **20**(24), 16485–16492 (2018). https://doi.org/10.1039/c8cp02663g
72. Song, F., Bai, L., Moysiadou, A., Lee, S., Hu, C., Liardet, L., Hu, X.: Transition metal oxides as electrocatalysts for the oxygen evolution reaction in alkaline solutions: an application-inspired renaissance. J. Am. Chem. Soc. **140**(25), 7748–7759 (2018). https://doi.org/10.1021/jacs.8b04546
73. Sookhakian, M., Ullah, H., Mat Teridi, M.A., Tong, G.B., Basirun, W.J., Alias, Y.: Boron-doped graphene-supported manganese oxide nanotubes as an efficient non-metal catalyst for the oxygen reduction reaction. Sustain. Energy Fuels **4**(2), 737–749 (2020). https://doi.org/10.1039/c9se00775j
74. Drummond, R., Huang, C., Grant, P.S., Duncan, S.R.: Overcoming diffusion limitations in supercapacitors using layered electrodes. J. Power Sources 433 (2019). https://doi.org/10.1016/j.jpowsour.2019.04.107
75. Garcés, L., Mendoza, R., Oliva, A.I., Garcia, C.R., Medina-Velazquez, D.Y., Oliva, J.: A sustainable and foldable supercapacitor made with electrodes of recycled soda-label/graphene/ZnO:Ca and its mechanism for the charge storage. J. Energy Storage **51** (2022). https://doi.org/10.1016/j.est.2022.104601
76. Abdi, Y., Mazaheri, A., Hajibaba, S., Darbari, S., Rezvani, S.J., Cicco, A.D., Saliba, M.: A two-dimensional borophene supercapacitor. ACS Mater. Lett. **4**(10), 1929–1936 (2022). https://doi.org/10.1021/acsmaterialslett.2c00475
77. Joshi, A., Tomar, A.K., Singh, G., Sharma, R.K.: Engineering oxygen defects in the boron nanosheet for stabilizing complex bonding structure: an approach for high-performance supercapacitor. Chem. Eng. J. **407** (2021). https://doi.org/10.1016/j.cej.2020.127122
78. Wu, R., Xu, H., Zhao, Y., Zha, C., Deng, J., Zhang, C., Huang, W.: Borophene-like boron subunits-inserted molybdenum framework of MoB2 enables stable and quick-acting Li2S6-based lithium-sulfur batteries. Energy Storage Mater. **32**, 216–224 (2020). https://doi.org/10.1016/j.ensm.2020.07.040

79. Rajagopal, S., Pulapparambil Vallikkattil, R., Mohamed Ibrahim, M., Velev, D.G.: Electrode materials for supercapacitors in hybrid electric vehicles: challenges and current progress. Condens. Matter. MDPI (2022). https://doi.org/10.3390/condmat7010006
80. Zhai, Z., Zhang, L., Du, T., Ren, B., Xu, Y., Wang, S., Liu, Z.: A review of carbon materials for supercapacitors. In: Materials and Design. Elsevier Ltd. (2022). https://doi.org/10.1016/j.matdes.2022.111017
81. Gao, T., Gong, L.J., Wang, Z., Yang, Z.K., Pan, W., He, L., Xu, W.: Boron nitride/reduced graphene oxide nanocomposites as supercapacitors electrodes. Mater. Lett. **159**, 54–57 (2015). https://doi.org/10.1016/j.matlet.2015.06.072
82. Jiang, H.R., Shyy, W., Liu, M., Wei, L., Wu, M.C., Zhao, T.S.: Boron phosphide monolayer as a potential anode material for alkali metal-based batteries. J. Mater. Chem. A **5**(2), 672–679 (2017). https://doi.org/10.1039/c6ta09264k
83. Bahadur, R., Singh, G., Li, M., Chu, D., Yi, J., Karakoti, A., Vinu, A.: BCN nanostructures conjugated nanoporous carbon with oxygenated surface and high specific surface area for enhanced CO2 capture and supercapacitance. Chem. Eng. J. **460** (2023). https://doi.org/10.1016/j.cej.2023.141793
84. Süleyman, E., De Wijs, G.A., Brocks, G.: DFT study of planar boron sheets: a new template for hydrogen storage. J. Phys. Chem. C **113**(43), 18962–18967 (2009). https://doi.org/10.1021/jp9077079
85. Li, L., Zhang, H., Cheng, X.: The high hydrogen storage capacities of Li-decorated borophene. Comput. Mater. Sci. **137**, 119–124 (2017). https://doi.org/10.1016/j.commatsci.2017.05.032
86. Rao, D., Zhang, L., Meng, Z., Zhang, X., Wang, Y., Qiao, G., Lu, R.: Ultrahigh energy storage and ultrafast ion diffusion in borophene-based anodes for rechargeable metal ion batteries. J. Mater. Chem. A **5**(5), 2328–2338 (2017). https://doi.org/10.1039/c6ta09730h
87. Hu, J., Zhong, C., Wu, W., Liu, N., Liu, Y., Yang, S.A., Ouyang, C.: 2D honeycomb borophene oxide: a promising anode material offering super high capacity for Li/Na-ion batteries. J. Phys. Condens. Matter **32**(6) (2020). https://doi.org/10.1088/1361-648X/ab4f4d
88. Mortazavi, B., Rahaman, O., Ahzi, S., Rabczuk, T.: Flat borophene films as anode materials for Mg, Na or Li-ion batteries with ultra high capacities: a first-principles study. Appl. Mater. Today **8**, 60–67 (2017). https://doi.org/10.1016/j.apmt.2017.04.010
89. Haldar, S., Mukherjee, S., Singh, C.V.: Hydrogen storage in Li, Na and Ca decorated and defective borophene: a first principles study. RSC Adv. **8**(37), 20748–20757 (2018). https://doi.org/10.1039/c7ra12512g
90. Habibi, P., Vlugt, T.J.H., Dey, P., Moultos, O.A.: Reversible hydrogen storage in metal-decorated honeycomb borophene oxide. ACS Appl. Mater. Interfaces **13**(36), 43233–43240 (2021). https://doi.org/10.1021/acsami.1c09865
91. Tang, X., Gu, Y., Kou, L.: Theoretical investigation of calcium-decorated B12 boron sheet for hydrogen storage. Chem. Phys. Lett. **695**, 211–215 (2018). https://doi.org/10.1016/j.cplett.2018.02.022
92. Baraiya, B.A., Som, N.N., Mankad, V., Wu, G., Wang, J., Jha, P.K.: Nitrogen-decorated borophene: An empowering contestant for hydrogen storage. Appl. Surf. Sci. **527** (2020). https://doi.org/10.1016/j.apsusc.2020.146852

Borophene-Based Materials for Gas and Wearble Sensing Applications

Sumit Kumar, Shivam Kashyap, Usama Ansari, Ankita Singh, Anjana Sarkar, and Deepak Poddar

Abstract Due to industrialization and urbanization, the degree of contamination of natural resources rises because of increasing concentration of pollutants in the environment that cause severe global issues like global warming, acid rain, and increased toxicity in air that affect human health in terms of severe or permanent neurasthenia syndrome, and chronic respiratory inflammation which reduce the average lifespan of human which makes the monitoring the concentration of health hazarding pollutant essential. As mono elemental 2D layered nanomaterial such as graphene is attentively used in sensors and biosensors, but due to lack of air stability and more susceptibility to oxidation, an alternative borophene mono elemental monolayer-based materials came into the limelight in 2015. Borophene is the crystalline atomic 2D monolayer sheet of boron which attracted immense attention due to its excellent air stability, high carrier mobility, photon-mediated superconductivity, high surface-to-volume ratio, high selectivity, mechanical compliance, flexibility, negative Poisson's ratio, defect sites, fast response, optical transparency, high sensitivity, ultrahigh thermal conductance and extensive detection range of 200 parts per billion to 100 parts per million which make it an excellent material for sensing application such as gas sensing, electrochemical sensing, excitonic and molecular sensing, etc. In this chapter, we will begin with the fundamental properties of borophene and the working mechanism of borophene in sensing applications in terms of both theoretical and experimental aspects. After that, we will delve into borophene-based nanocomposite in sensing applications such as gas sensing, chemosensing, biosensing, and molecular and excitonic sensing.

Keywords 2D materials · Borophene · Gas sensing · Chemosensor · Biosensor · Monoelemental layer · Chemically modified borophene

---

S. Kumar · S. Kashyap · U. Ansari · A. Singh · A. Sarkar · D. Poddar (✉)
Department of Chemistry, Netaji Subhas University of Technology (N.S.U.T), Azad Hind Fauj Marg, Sector-3, Dwarka, New Delhi 110078, India
e-mail: Poddardeepakpoddar@gmail.com

R. Khan et al. (eds.), *2D Boron Nanosheets*, Engineering Materials,
https://doi.org/10.1007/978-981-97-7014-4_7

1 Introduction

Since the discovery of borophene in 2015, it has garnered significant attention in sensing applications, owing to its extraordinary properties and potential advantages over other traditional sensing materials [1]. Borophene is a 2D material consisting of boron atoms arranged in a honeycomb lattice structure exhibiting unique characteristics that make it a potential material for sensing applications. The properties that set borophene apart from other materials are a high surface-to-volume ratio and large specific surface area that contribute to enhanced sensing capabilities compared to bulk materials. It also makes borophene highly sensitive to external stimuli such as analytes and gas [2]. Moreover, borophene exhibits excellent mechanical flexibility that allows its integration into flexible and wearable sensors, which opens new avenues for real-time monitoring and healthcare applications. Furthermore, borophene stability in harsh environments is a crucial advantage for sensing applications, which allows it to withstand high temperatures, humidity, and mechanical deformation without significant degradation in sensing performance. This stability ensures the long-term reliability and durability of borophene-based sensors, making them suitable for use in challenging conditions. One of the limitations encountered with traditional materials is their sensitivity and selectivity. In contrast, borophene demonstrates excellent sensitivity and selectivity towards various analytes, volatile organic compounds (VOC), and toxic gases [3]. The selectivity arises from a unique electronic structure and fast charge transfer capabilities that enable it to be selective to absorb specific gas molecules and exhibit distinct response patterns. The tailorable selectivity is crucial for accurately detecting and monitoring environmental pollutants, healthcare diagnostics, and industrial safety.

This chapter provides a comprehensive overview of sensing applications of borophene-based materials, as shown in Figure 1 [4]. By synthesis and characterization of borophene, we increase our understanding of the sensing mechanisms and explore strategies for enhanced sensing performance. The chapter seeks to address the growing interest in borophene-based materials for various sensing applications that shed light on the unique properties of two-dimensional borophene and discuss its advantages over other sensing materials. Additionally, we intend to highlight the future perspectives and challenges to stimulate further research in this rapidly evolving field.

1.1 Properties of Borophene for Sensing Application

Due to the 2D honeycomb structure of borophene, it possesses unique properties, making it a preferred material over other traditional materials. Some of the properties of borophene are listed below.

1. *High surface-to-volume ratio*:—Borophene exhibits ultrathin sheet structure, resulting in a high surface-to-volume ratio, which allows for a larger surface

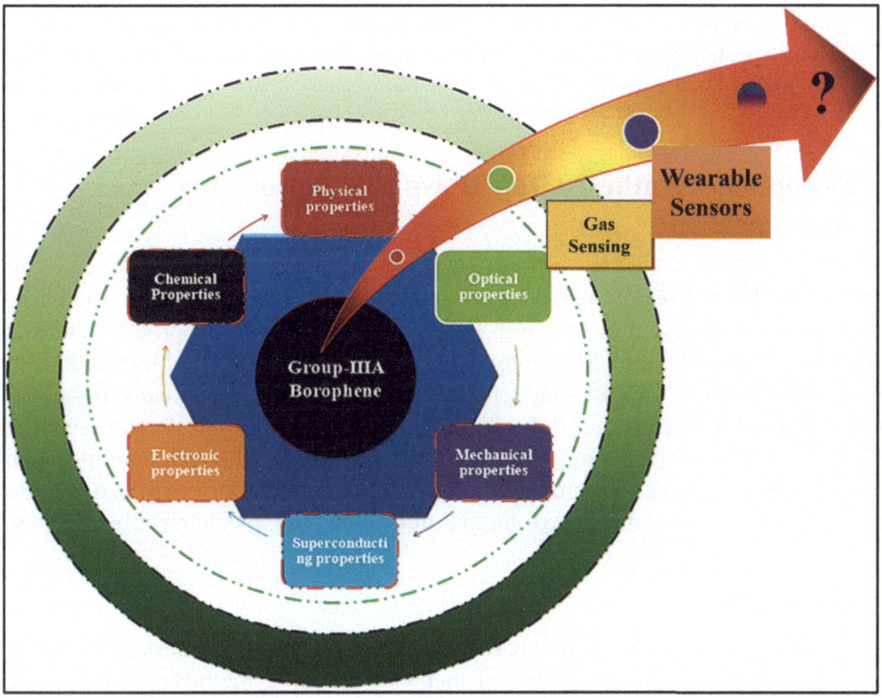

Fig. 1 Overview of the chapter and properties and applications of borophene-based 2D nanosheets (Reprint with permission) [4]

area available to interact with analytes, enabling enhanced sensitivity in sensing applications or increased surface-to-volume ratio leads to enhanced chances of analyte get adsorb, thus improve the sensing performance [5].

2. *Large specific surface area:*—the 2D structure of borophene provides it with a large specific surface area, which facilitates increased adsorption and interaction with targeting molecules which helps in detecting trace amounts of analytes, leading to improved sensitivity of borophene-based sensors.
3. *Selective adsorption:*—Borophene exhibits high affinity towards certain analytes such as NO, NO_2, and NH_3 [6] allowing it for enhanced selectivity in sensing applications. The affinity of borophene towards different analytes can be fine-tuned by chemical modification which enables the detection of specific analytes with high precision.
4. *Mechanical Flexibility:*—Borophene possesses significant mechanical flexibility that enables its integration into wearable and flexible sensors. Due to mechanical flexibility, borophene-based sensors can withstand mechanical deformation and can be used for real-time monitoring of physiological parameters and environmental conditions [5].
5. *Plasmonic effects:*—Borophene exhibits plasmonic effects [7], which open new avenues to be utilized in ultrasensitive sensing applications. Plasmons enable

matter-light interaction, leading to enhanced sensitivity in Surface Enhanced Raman Spectroscopy (SERS) that allows for highly sensitive detection of minute concentrations of analytes.

2 Borophene Synthesis: Top-Down and Bottom-Up Approach

Two approaches are widely opted for the synthesis of 2D nanomaterials, including bottom-up and top-down as shown in Figure 2. The bottom-up approach begins at the atomic scale and uses various physical, chemical, and biological interactions to obtain nanoparticles. This approach includes methods such as physical vapor deposition (PVD), Chemical vapor deposition (CVD), and wet chemical synthesis. The top-down approach uses energy extracted from biological, chemical, and physical sources to break complex structures into simpler ones. Methods used in the top-down approach are ion-intercalation exfoliation, ultrasonication, mechanical breakdown, and etching methods [8].

Bottom-up approach

Bottom-up synthesis includes mainly two routes to fabricate nanomaterials, including wet vapor methods and wet chemical methods. In vapor methods, experimental synthesis of 2D borophene was carried out by CVD, molecular beam (MBE), and electron beam epitaxy (EBE) method on various substrates such as Au (111), Al (111), Ru (001), Ir (111) and quartz. An atomically thin 2D γ-boron film over copper foil was fabricated by *Tai et al* using the CVD method, boron and boron oxide powder were used as precursors for the boron source and H_2 as carrier gas. Wet chemical methods involve the use of solvents and chemical reactions cause the exfoliation of boron flakes or clusters into borophene sheets. One of the most common wet chemical methods is sonochemical synthesis. In this method, ultrasonic waves are used to create shockwaves that break down the bonds between boron atoms and result in borophene sheets. Another example of a wet chemical method is liquid-phase exfoliation, in which a solvent is used which intercalates between the boron atoms and causes weakening of bonding between boron atoms in the bulk. Then boron flakes are

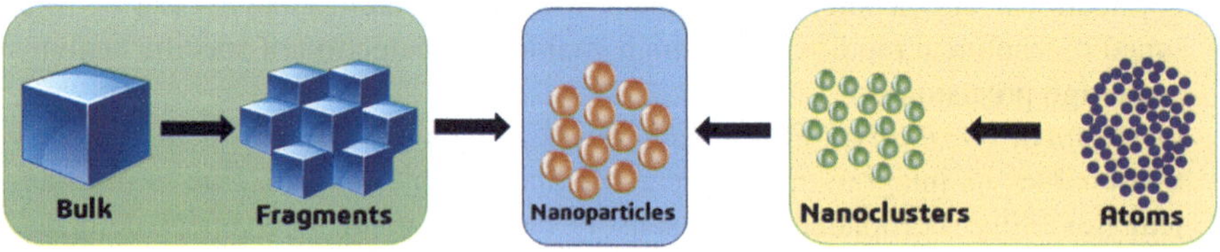

Fig. 2 Nanomaterial synthesis by a top-down and bottom-up approach. Reproduced with permission from [9]

sonicated in the solvent, attributed to exfoliated 2D borophene nanosheet structure [10].

Top-down approach

The top-down approach typically begins with bulk boron material, which is thinned down to obtain desired nanosheets. Top-down offers various advantages over bottom-up, including scalability, control over size and thickness, and the ability to create large-area films. One of the most employed methods in the top-down approach is mechanical exfoliation. In this method, the bulk material is mechanically cleaved and broken down to obtain thin flakes and nanosheets. This technique relies on weak Vander Waal forces between atom layers, allowing for the separation of individual layers through scotch tape peeling or mechanical shearing [11].

3 Why Borophene Over Other 2D Nanomaterials?

The unique properties of 2D materials, such as boron nitride (BN), graphene, tungsten disulfide (WS_2), molybdenum disulfide (MoS_2), and phosphorene, have attracted attention for futuristic nanodevice applications, including sensing. They offer large surface areas and enhanced physical and chemical properties due to their 2D nature. However, certain drawbacks hinder their utilization in detecting toxic gases, primarily the lack of an electronic bandgap in materials such as graphene, which reduces the sensitivity and selectivity [2].

To overcome these drawbacks, researchers have searched for alternative 2D materials having higher mobilities or comparable properties to graphene and superior sensing capabilities. One such material is borophene, characterized by its metallic structure and increased electronic density along ridge lines. Borophene exhibits a high-performance potential for sensing applications, as demonstrated in a study conducted by Ranjan et al., where they used free-standing borophene sheets to detect ammonia [12].

Extensive investigations have been conducted by Huang et al., Cui et al., Shukla et al., Liu et al., and Omidvar to explore the interactions, adsorptions, and sensing properties of poisonous gases on borophene when supported by various substrates. These investigations were carried out using density functional theory (DFT) calculations [13].

Shukla and colleagues utilized DFT computations in their study, employing the Vienna Ab initio Simulation Package (VASP) software. The Perdew–Burke–Ernzerhof exchange–correlation function was adopted, and van der Waals corrections were incorporated using the Grimme (DFT-D3) approach. The examination of the electrical and transport characteristics of monolayer borophene sheets yielded notable binding energies for all examined gases, except carbon dioxide (CO_2). The investigation into the electrical and transport properties of monolayer borophene sheets revealed significant binding energies for all gases studied, except for carbon dioxide (CO_2) [14].

Liu et al. conducted DFT calculations using the Cambridge Sequential Total Energy Package (CASTEP) in Materials Studio 6.1 and demonstrated that borophene sheets can act as electron donors. This property enables easy interaction and chemical bonding with electron-acceptor molecules. According to their theoretical calculations, borophene exhibits excellent sensing capabilities specifically for CO_2 gas molecules.

4 Sensing Mechanism of 2D Nanosheets

Sensing behavior in 2D layered nanomaterials has been extensively studied, and different explanations have been proposed in scientific literature. The prevailing theory suggests that gas molecules physically adsorb on the surface of nanosheets due to factors such as long adsorption distances, weak adsorption energy, minimal charge transfer, and nearly constant electronic structure [15]. Nevertheless, the discoveries primarily relied on meticulously constructed two-dimensional nanosheets with a single layer, which possess limited sensitivity owing to their unreactive surface. However, it is important to note that defects are unavoidably introduced during the synthesis of the material, and these imperfections tend to be more chemically active compared to the unblemished lattice sites. As a result, there is a phenomenon known as chemical adsorption that takes place when reactive gas molecules interact with the faulty or modified sites present over the top of 2D nanosheets.

DFT calculations have provided insights indicating that gas molecules can be adsorbed at distances that are less than the combined total of two atomic radii. This phenomenon leads to a notable increase in adsorption energy and a more substantial charge transfer than physical adsorption. The powerful hybridization between the molecular orbitals of the analyte and the 2D nanosheet results in promising modifications to their electronic characteristics. This intense contact is responsible for the observed changes. A process of charge transfer occurs between the chemically reactive gas and the nanosheets with a two-dimensional layered structure, resulting in a modification of the sensing materials' electrical conductivity or resistivity properties.

Two primary charge exchange mechanisms have been widely proposed:

(i) The exchange of charge is contingent upon the relative orientations of the HOMO and LUMO of the gas molecules concerning the fermi level of the pristine material. If the HOMO exceeds the fermi point, a charge transfers toward the two-dimensional nanomaterials. Conversely, when the LUMO falls below the fermi point, charge is extracted from the nanosheet and directed towards the analyte molecule [16].

(ii) The process of electronic charge exchange between analyte and 2D nanosheets is additionally modulated by the hybridization occurring between the HOMO and LUMO, as well as the orbitals of the nanosheets. The level of mixing is influenced by the overlapping of interacting orbitals and the reciprocal of their energy difference. When sensing devices are open to particular reactive gas molecules, these molecules become adsorbed onto the surface of the two-dimensional sensing materials. These materials function as either a donation of electrons or an acceptance of electrons. By deliberately enhancing the affinity of the charge towards the sensing materials, it becomes feasible to augment the sensitivity and specificity of the gas sensors. Conversely, if the sensors come into contact with air or nitrogen, it is anticipated that the gas molecules will undergo desorption from the surface of the sensing materials. The enhancement of desorption reversibility in gas sensors can be achieved by using measures to prevent excessive charge transfer. Hence, with the manipulation of the charge affinity of the gas species, it becomes possible to exert control over the properties of the gas-sensing materials [17].

5 Theoretical Modeling of Sensing Mechanism of Borophene-Based 2D Nanosheets

Several theoretical models and computational simulations are used to explain and predict the sensing behavior of borophene-based 2D nanosheets. These models shed light on the interaction energies, charge transfer mechanisms, and changes in electronic characteristics caused by gas adsorption on the borophene surface.

DFT computations are commonly used to investigate different chemical systems' electrical structure and adsorption properties. DFT gives a quantum mechanical description of the system, allowing for the calculation of adsorption energies, charge transfer, and electronic property changes. In addition, modern computational approaches like molecular dynamics simulations and Monte Carlo simulations can all be used to explore both the thermodynamic and kinetic aspects of gas adsorption on 2D nanosheet surfaces. These simulations can provide information regarding gas diffusion, desorption, and gas species stability. Muhammad Isa Khan et al. studied the theoretically studied adsorption behavior of different gas molecules like CO, NO, CO_2, NO_2, and NH_3 at the surface of borophene with the help of DFT under van der Waals interaction, results obtained during studies are represented in Fig. 3 [18]. Researchers want to understand borophene gas sensing mechanisms thoroughly by combining theoretical modeling and experimental validation, allowing for the creation and optimization of susceptible and selective borophene-based sensors [18].

6 Application of Borophene Based Materials

6.1 Gas Sensing Applications

The extremely thin nature of the borophene nanosheets results in extraordinary features that provide various advantages for gas sensing mechanisms. The increased surface area gives more active places for gas molecules to interact with the material,

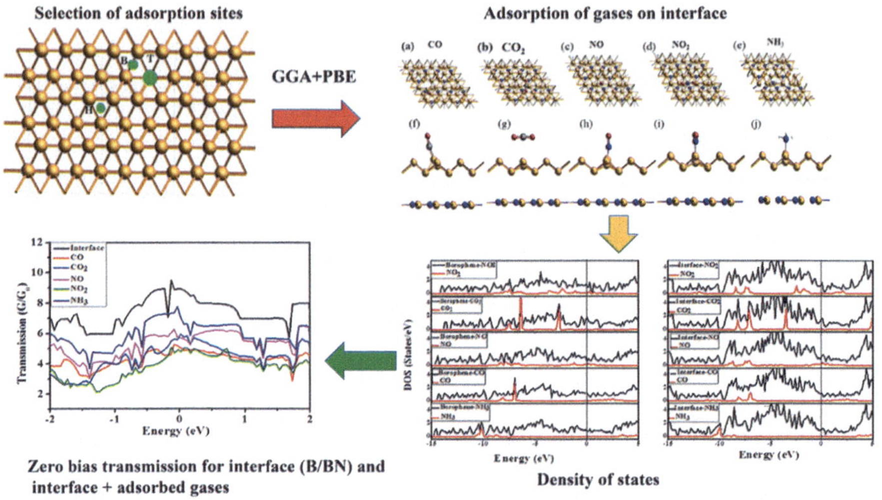

Fig. 3 Illustration of the adsorption site, top and side interface between gases and borophene, their density of state, and their zero-transmission interface and gas adsorption. (Reprint with Permission) [18]

resulting in improved sensitivity. Because of this increased sensitivity, borophene nanosheets are well-suited for applications requiring high sensitivity, such as environmental monitoring of contaminants or tracer gas detection in industrial processes [19].

The characteristics of the gas molecule and the surface of the borophene determine the kind of adsorption that takes place between them. In comparison to non-polar gas molecules like CO_2, polar gas molecules like CO and NO_2 are more prone to experience chemical adsorption. The adsorption procedure involves the borophene surface as well. For instance, borophene with a high defect density has a higher propensity to be adsorbed chemically than borophene with a low defect density [13].

Borophene-based gas sensors work by adsorbing gas molecules and then transferring charge. Gas molecules interact with the borophene surface by donating or taking electrons that alter the electrical conductivity of the borophene film. This change in conductivity can be monitored and correlated with the target gas concentration [20].

6.1.1 Interaction Mechanism of Gas Analyte with Borophene

When using borophene-based gas sensors, the interaction between the gas molecule and borophene is crucial to the sensing process. Numerous factors, including the type of gas molecules, the surface functionalization of the borophene, and the borophene's structural characteristics, affect the interaction mechanism. It is possible to observe several typical interaction mechanisms in borophene-based gas sensors. The following are these mechanisms [21].

Charge Transfer: Both the gaseous molecules and the borophene exchange electrons, changing the charges of both. By binding or donating electrons from the surface of borophene, gas molecules can alter the concentration and conductivity of the charge carriers. Both chemisorption and physisorption phenomena exhibit the previously mentioned mechanism, with the degree of contact being the determining factor in their relationship. Many gases, such as sulfur dioxide (SO_2), hydrogen (H_2), and carbon monoxide (CO), exhibit charge transfer interactions with borophene [22].

Physisorption: The weak van der Waals forces between gas molecules and the borophene surface are called physisorption. Using this approach, the gas molecules have negligible contact with the borophene surface, preventing any significant chemical reactions. When physisorbed onto borophene, gas molecules can alter the material's charge carrier concentration, electrical properties, and conductivity. Nitrogen (N_2), oxygen (O_2), and carbon dioxide (CO_2) are only some examples of gases that frequently demonstrate this behavior [23].

Chemisorption: The interaction between gas molecules and borophene is significantly enhanced and exhibits more chemical relevance during the process of chemisorption. The electrical structure and conductivity of borophene are modified through the establishment of chemical bonds between gas molecules and the surface atoms, hence inducing alterations in its properties. Chemisorption is a commonly seen phenomenon in which gases such as ammonia (NH_3), hydrogen sulfide (H_2S), and nitrogen dioxide (NO_2) undergo the process of adsorption, as shown in Fig. 4 [24].

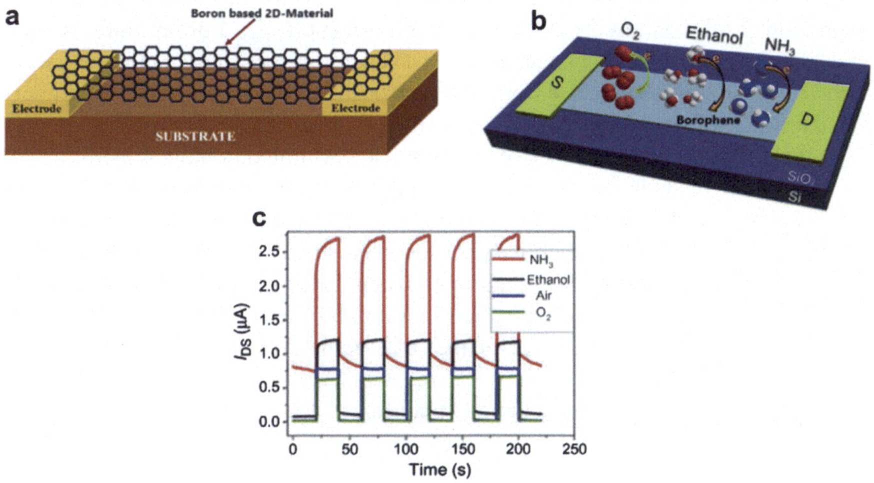

Fig. 4 Figure **a** represents the sensing device consisting of boron-based electrode material, Figure **b** illustrates the sensing of O_2, Ethanol, and NH_3, and **c** shows the response generated by the device toward gas sensing. (Reprint with the permission) [24]

Surface Doping: Surface doping is a process that alters the concentration of charge carriers on the surface of borophene by adsorbing gas molecules. The conductivity and electrical properties of borophene can be modified due to the presence of adsorbed gas molecules, which function as dopants by either donating or absorbing charge carriers. The aforementioned mechanism is commonly observed in the context of gaseous substances such as nitrogen dioxide (NO_2) and carbon monoxide (CO) [13].

It's important to remember that the precise contact process can be affected by a wide variety of factors, such as pressure, humidity, temperature, and defects or functional groups on the borophene surface. Borophene-based gas sensors can improve their sensitivity and selectivity by doping, inserting functional groups, or adding defects to the surface, all of which modify the interaction process.

Designing specialized gas sensors, optimizing sensor performance, and modifying borophene's surface properties for improved gas sensing applications all require an understanding of the interaction process between gas molecules and borophene. Researchers frequently use experimental experiments, theoretical calculations, and computational simulations to look into and characterize these interaction mechanisms in borophene-based gas sensors.

In their recent study, Chuang Hou, Guoan Tai, Yi Liu, and Xiang Liu (2022) conducted an empirical investigation on the chemical detecting capabilities of NO_2 using a gas sensor based on borophene. The experiment was conducted at room temperature, and the results have shown that the borophene sensor exhibited a broad detection range spanning from 0.2 ppm to 100 ppm for NO_2 gas. The borophene sensor can attain a significant sensitivity of up to 445% when exposed to a gas concentration of 100 parts per million (ppm), and it exhibits a low experimental limit of detection (LOD) as low as 200 parts per billion (ppb) [3]. Furthermore, Ranjan et al. (2019) opted for boron nanosheets as the sensing layer. These nanosheets were obtained by the sonochemical exfoliation approach. Subsequently, the researchers utilized a straightforward spin-coating technique to construct an ammonia gas sensor on a sodium silicate substrate. The resulting sensor exhibited notable sensitivity and selectivity [12]. In their investigation, Muhammad Sajjad and Peter Feng (2014) effectively employed the CO_2-PLD technique to synthesize boron nitride nanosheets. They investigated the surface chemistry of these nanosheets and proposed their potential suitability for facilitating the reaction between CH_4 gas molecules and B and N atoms. This reactivity ultimately resulted in a notable sensitivity towards CH4 gas molecules, indicating the potential for the development of a methane-sensing gas sensor device [25] (Table 1).

Here we discuss some characteristic features of borophene and its based 2D nanosheets related to their application in gas sensing:

1. Adsorption properties:—The ability of borophene to adsorb gas molecules on its surface is a result of its powerful adsorption properties. The electronic structure of the local strain of borophene may change as a result of the adsorbed gas molecules, which might result in observable changes in its characteristics that can be measured and observed.

Table 1 The adsorption energy (E_{ad} in joule), the minimum distance between analyte and borophene, and the Bader charge on the molecule of the most stable configurations were shown with the help of chemical/physical adsorption reprint with permission [23]

Gas molecule	Adsorption type	E_{ad} (joule)	Distance (nm)	Charge(e)
CO	Chemical	-1.84×10^{-19}	0.15	-41×10^{-2}
CO_2	Physical	-1.02×10^{-19}	0.29	-4×10^{-2}
NO	Chemical	-2.35×10^{-19}	0.14	-124×10^{-2}
NO_2	Chemical	-3.39×10^{-19}	0.15	-80×10^{-2}
NH_3	Chemical	-2.43×10^{-19}	0.16	12×10^{-2}

2. Selectivity:—By adding various functional groups or metal atoms to the surface of the borophene, its gas-detecting abilities can be improved even more. The capacity of borophene to differentiate between various analytes can be improved by altering the surface so that it interacts only with particular gases or gas molecules [2].
3. Sensitivity:—A borophene-based sensor's sensitivity refers to its capacity to identify and measure the presence of a target gas. Due to its high surface-to-volume ratio and tunable electrical characteristics, borophene demonstrates great sensitivity to a variety of gases. The electrical conductivity or optical characteristics of borophene can be significantly altered by changes in the gas environment, enabling sensitive detection. Higher sensitivity means the sensor device is capable of detecting the possible lower concentration of targeted gas molecules.
4. Detection range:—The minimum detection range in terms of concentration or amount of the gas molecule detected by the sensors is known as the detection limit. A wide variety of gases, such as harmful and dangerous gases like ammonia, nitrogen dioxide, carbon monoxide, and volatile organic compounds (VOCs), have been successfully detected using borophene. By modifying the borophene's composition, doping, or functionalization, its sensitivity and selectivity can be altered. The lower detection limit represents the more sensitive device.
5. Rapid response and recovery:—When a sensor is exposed to a target gas, its response time is the amount of time it takes for the output signal to stabilize, whereas the time taken by the sensor's output signal to return to its baseline as soon as the target gas is eliminated or its concentration becomes lower than the detection range it is known as recovery time. Real-time monitoring of gas concentrations is now possible because of the quick reaction and recovery times of borophene-based gas sensors. This speedy reaction is owing to the gas molecules' effective interaction with borophene, which enables prompt detection and subsequent desorption.

6.2 Application of Borophene-Based Gas Sensors in Environmental Monitoring, Industrial Safety, and Healthcare Sector

Environmental Monitoring:—Borophene, as a gas sensor, may be used for environmental monitoring in a variety of applications. Among the potential applications of borophene-based gas sensors for environmental monitoring are:

1. Air Quality Monitoring:—Air pollutants including volatile organic compounds (VOCs), nitrogen oxides (NOx), Sulphur dioxide (SO_2), carbon monoxide (CO), and several other dangerous gases may be monitored using borophene-based gas sensors. The concentration of harmful contaminants in the air may be detected and measured with the use of these sensors, allowing authorities to take the necessary action to improve air quality management.
2. Environmental Leak Detection:—Toxic or flammable gas leaks from industrial facilities, chemical storage sites, or pipelines can be found using borophene sensors. It is essential to find these leaks quickly to avoid potential risks and environmental degradation [26].
3. Indoor Air Quality:—Human health and comfort may be dramatically impacted by indoor air quality. To monitor indoor air pollutants and give prompt alerts or ventilation advice for maintaining a healthy atmosphere, borophene-based gas sensors can be used in interior locations such as homes, workplaces, and public spaces.
4. Industrial Emission:—During their operations, industries frequently produce harmful gases and pollutants. To continually check for pollutants and guarantee environmental rules are being followed, emission control systems might use borophene-based gas sensors.
5. Soil Contamination Monitoring:—The ecosystems and groundwater in the area may be at risk from certain gases released from polluted soil. By detecting and measuring these gases, borophene-based gas sensors can help determine the extent of soil pollution and help execute remediation strategies [27].

Industrial Safety:—These gas sensors have the potential to significantly improve workplace safety by constantly detecting the presence of dangerous gases and sending out instantaneous notifications to stop mishaps and safeguard employees. Here are a few particular uses for borophene-based gas sensors in the field of industrial safety:

1. Combustible Gas Monitoring:—Gas sensors based on borophene can identify flammable gases including hydrogen (H_2), propane (C_3H_8), and methane (CH_4). To reduce the risk of fire or explosion in industrial settings, it is essential to maintain acceptable levels of flammable gases [28].
2. Toxic Gas Detection:—Toxic gases including hydrogen sulfide (H_2S), ammonia (NH_3), chlorine (Cl_2), and many others frequently encountered in industrial environments may be detected with borophene sensors. Preventing worker exposure and related health risks is possible with early identification of these gases.

3. Workers' safety:—Borophene-based gas sensors can be found in respirators and gas masks, among other personal protective equipment, to provide users with real-time information on gas concentrations and help them respond appropriately to changing circumstances.
4. Process Monitoring and Control:—Industrial process control systems can use borophene sensors to monitor gas concentrations and ensure that processes are operating within safe bounds. If gas levels exceed safety limits, emergency shutdowns or control changes can be initiated to prevent incidents.

Healthcare Sector:—Borophene-based gas sensors have tremendous potential for applications in the healthcare sector due to their sensitivity, selectivity, and ability to detect a wide spectrum of gases [29]. Here are some potential applications for borophene-based gas sensors in healthcare:

1. Real-time Diagnosis:—Gas sensors based on borophene that are portable and can be used in distant or resource-constrained environments are a possibility. For certain disorders, they might offer quick, affordable diagnoses, allowing for early intervention and better patient outcomes.
2. Breath Analysis and Anesthesia Monitoring:—Borophene sensors can detect traces of gases suggestive of metabolic problems or respiratory diseases in a person's breath. This non-invasive method could aid in the early detection of disease and the development of personalized treatment programs; on the other hand, borophene sensors can be integrated into anesthesia delivery systems during surgery and other medical operations to continually monitor the concentration of anesthetic gases, ensuring the patient receives the correct dose and lowering the risk of complications [30].
3. Patient Monitoring and Elderly Care:—The integration of gas sensors based on borophene into wearable devices has the potential to enable the monitoring of a patient's breath or body odor to detect early indications of infections or metabolic imbalances. Borophene sensors could be utilized to monitor indoor air quality in senior housing developments and nursing homes, looking for potentially dangerous chemicals like harmful gases. This would offer healthcare practitioners vital data to make necessary adjustments to treatment programs.
4. Environmental Monitoring in Healthcare Facilities:—Healthcare institutions are susceptible to incidents involving chemical spills or gas leaks. Borophene-based sensors provide the capability to consistently monitor the quality of indoor air and promptly notify personnel in the event of the emission of dangerous gases.
5. Ensuring the reliability, precision, and calibration of these sensors is crucial to optimize their efficacy across diverse applications. Regular maintenance and calibration are crucial in ensuring the optimal performance and accuracy of systems in critical situations. Moreover, the incorporation of these sensors into integrated safety systems can offer a multifaceted strategy for industrial and health safety, thereby augmenting the overall safety and welfare of workers and the surrounding environment [31].

6.3 Mechanical Flexibility and Wearable Sensors

Borophene's mechanical flexibility is greatly influenced by its 2D structure and atomic thinness. As it is just a single atom thick and is very light, it can bend and adapt to curved surfaces without suffering substantial damage. It is remarkably strong mechanically while being thin, allowing it to sustain mechanical strain and stress through prolonged usage. When it comes to wearable sensors, mechanical flexibility is an extremely important factor, since these sensors need to conform to the contours and motions of the human body to function properly [32].

Borophene has anisotropic mechanical characteristics, which is another benefit of using this material. This indicates that its mechanical behavior changes depending on the direction in which it is deformed, providing researchers with the ability to modify its flexibility according to particular designs for wearable sensors. Borophene's mechanical performance may be optimized to suit the criteria that are needed if one controls the development and orientation of the crystals that make up the material. The surface of borophene may be functionalized or easily combined with other materials, such as flexible substrates or polymers, to further improve its mechanical characteristics. For researchers to increase the material's flexibility, stretchability, and general mechanical performance, they may either make chemical alterations or mix it with other materials. Borophene is a material that is attractive for wearable sensors because it permits effective data collection and transmission. In addition, borophene has great electrical conductivity, which, when combined with its mechanical flexibility, makes it a desired material [33]. Once the charge density and capacity of borophene-based devices have been efficiently adjusted, it is anticipated that the charging time of implantable medical devices (IMDs) might be substantially decreased to a matter of minutes. The implementation of a hybrid power source, combined with LiS, in β12 borophene, has resulted in the elimination of external batteries for IMDs. This integration has also led to improved charge and discharge intervals, ensuring faster operation [34].

For example, Chen et, al. used borophene/ecoflex nanocomposite to build a flexible nanocomposite for healthcare electronics, the findings reveal that the generated nanocomposite delivers constant output performance even after severe deformation treatments such as folding, stretching, twisting, and washing operations. Furthermore, the composite is paired with a smart keyboard configuration and a robotic system to conduct an upper-limb medical assistive interface [35]. Furthermore, Hou et al. fabricates a pressure-sensitive sensor for health monitoring using 2D borophene semiconducting sheets having specific high sensitivity around 2.16 kPa^{-1}, low detection limit i.e., 10 P, and great reproducibility with more than 1000 cycles. Their results show the sensor based on borophene may be employed as wearable electronics for health monitoring, speech recognition, and wireless human motion detection, as well as in smart robots for human–machine interaction [36]. In addition, P. K. Sharma et al. show the potential of borophene-based smart sensors for point of care (POC)

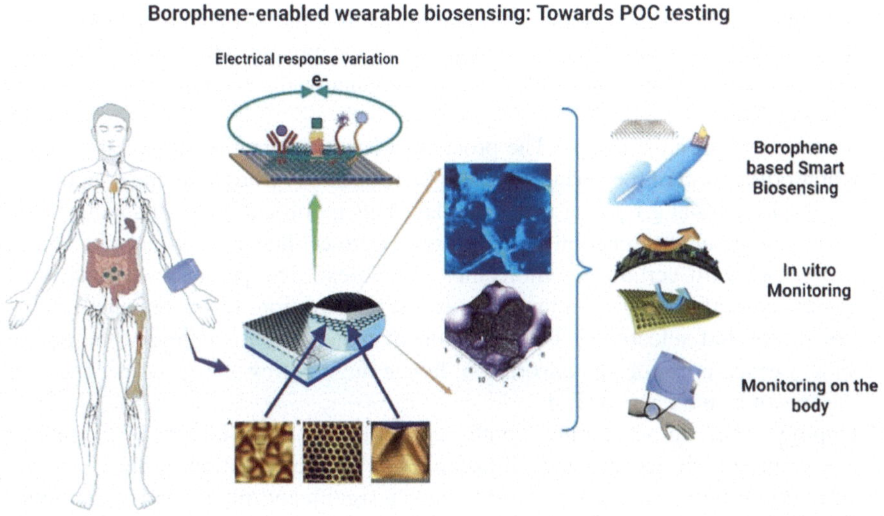

Fig. 5 Borophene-based wearable device and their sensing towards POC testing. (Reprint with permission) [29]

diagnosis and monitoring by mobile phone as shown in Fig. 5. They offer the functionalization of borophene by bioactive groups and deposition on flexible substrates are prominent factors for wearable smart-monitorable sensors [29].

Borophene has a lot of promise to be used in wearable sensors, but it's important to remember that research is still being done in this area, so it could be a while before it's widely used in commercial applications. Despite this, borophene has a lot of potential. Borophene is only one example of a kind of two-dimensional material that is always undergoing development, and scientists are constantly looking into new methods to improve and exploit the features of these materials for a wide range of applications, including wearable sensors. Borophene has the potential to bring about a sea change in the field of wearable technology, provided that it continues to make strides in scientific growth [37].

7 Strategies for Enhancing Sensing Performance

There are various methods to improve the sensing performance of borophene-based sensors:

1. Nanoparticle Integration:—The introduction of nanoparticles onto the surface of borophene can give rise to a phenomenon commonly referred to as the "nano-structuring effect." The nanoparticles function as catalysts, facilitating the processes of gas adsorption and reaction. This process increases the magnitude of the sensor's response signal and improves its ability to detect and measure

trace gases with greater sensitivity. The mutually beneficial interaction between the borophene substrate and the nanoparticles establishes a favorable setting for the interaction of gas molecules, hence enhancing the overall efficacy of gas sensing [38].

2. Surface Functionalization:—The process of functionalization entails the chemical modification of the surface of borophene sensors through the introduction of certain functional groups or nanoparticles. This modification results in a change to the chemical composition of the sensor's surface, hence improving its ability to interact with certain gas molecules of interest. The presence of functional groups can offer supplementary sites for gas adsorption, hence enhancing both sensitivity and selectivity. The inclusion of nanoparticles increases the accessible surface area for gas adsorption, hence augmenting the responsiveness of the sensor to trace gases [20].
3. Doping:—The act of doping entails the deliberate introduction of exogenous atoms into the lattice structure of borophene. The perturbation of the electronic structure of borophene is observed to have a significant impact on its sensitivity towards certain gases. The sensor's capacity to adsorb and interact with specific gases can be precisely adjusted by modifying the dopant type and concentration. Dopants can modify the charge distribution and electrical characteristics of borophene, thereby influencing its interaction with gas molecules and ultimately augmenting its gas-sensing capabilities [39].
4. Gas Preconcentration:—The process of gas preconcentration entails the concentration of molecules of trace gases before their interaction with the surface of the sensor. This phenomenon leads to an increased sensitivity of the sensor towards low gas concentrations since a greater number of gas molecules are present for interaction within a limited volume. Preconcentration processes encompass the utilization of specialized materials for adsorption or the implementation of microfluidic systems, hence enhancing the sensor's capacity to detect minuscule gas amounts [40].
5. Temperature Optimization:—The manipulation of the operational temperature of borophene sensors has the potential to exert an impact on the kinetics of gas adsorption as well as the mechanisms underlying the interaction process. The act of sensing at various temperatures can enhance gas desorption and reaction processes, leading to improvements in selectivity and response time. The technique of temperature modulation enables the manipulation of the surface energetics of the sensor, hence affecting its sensitivity towards various gases [41]. This particular method has the potential to enhance the performance of the sensor to achieve optimal results for specific gas targets.

In brief, the aforementioned scientific approaches utilize ideas derived from surface chemistry, nanomaterial engineering, electronic structure manipulation, and dynamic sensing to augment the sensitivity, selectivity, and overall efficacy of sensors based on borophene. By customizing these tactics to suit the specific demands of detection applications, researchers and engineers can enhance the efficiency of borophene sensors for a range of industries, encompassing healthcare, environmental monitoring, and industrial safety.

8 Challenges and Future Perspective

However, besides of unique mechanical, optical, and electronic properties of borophene-based sensors, there are still many challenges and limitations in the practical implementation. Some of these are:

1. Scalability: The current state of borophene synthesis is restricted by low yield, substrate dependence, and high temperature in methods like chemical vapor deposition and molecular beam epitaxy (MBE), which makes it difficult to produce and integrate borophene on a large scale for sensing devices [42]. Thus, scalable, substrate-independent, low-temperature synthesis techniques that can yield consistent, superior films are required.
2. Reproducibility: Temperature, pressure, and substrate temperature are among the synthesis parameters that have a significant impact on the electronic and structural characteristics of borophene. Furthermore, in natural environments, borophene is vulnerable to oxidation and degradation. These problems have an impact on the sensor's reproducibility, stability, and properties [43]. Thus, to guarantee consistent and dependable sensing performance, it is necessary to optimize the synthesis parameters and create an efficient protection plan for borophene.
3. Cost-effectiveness: The synthesis of borophenes requires expensive machinery, including electron guns, mass spectrometers, and ultra-high vacuum chambers. Moreover, the cost of the substrates—such as silver and copper—that are needed for borophene growth drives up the cost of both the fabrication of borophene-based sensors and the sensor itself. To cut costs, it is therefore necessary to identify less expensive equipment and substrate substitutes.

Future Research Directions

1. Borophene functionalization: Because of its diverse chemistry, borophene can be functionalized in a variety of ways with different atoms or molecules, changing its electrical structure and band gap while also introducing new properties including magnetism, catalysis, and biocompatibility.
2. Examining innovative structures: Borophene may develop a variety of architectures with distinct topologies and geometries, including nanosheets, nanoribbons, nanotubes, and nanoflowers, which all affect the surface area and transport characteristics of borophene. Additionally, these structures give rise to novel phenomena that influence the selectivity and performance of sensing, like quantum confinement and edge effects.
3. Borophene-based hybrid structures: To create heterostructures or hybrid systems, borophene can be combined with other 2D materials including graphene, hexagonal boron nitride (h-BN), and transition metal dichalcogenides (TMDCs). The

resultant hybrid systems have unique qualities and synergistic effects not seen in the constituent parts, which raises the sensitivity and sensing performance [44].

9 Conclusion

In conclusion, this chapter has given a thorough overview of the extraordinary promise shown by materials based on borophene in a variety of sensing applications. Because of its essential special mechanical, chemical, and electrical characteristics, borophene is a remarkable material for use in sensing applications. Borophene sensors possess unparalleled selectivity, sensitivity, and quick response times that could revolutionize the field of sensing technology. Pore-level pollution detection to real-time biomolecule monitoring: borophene-based sensors have the potential to improve our comprehension of intricate systems and spur innovation in a variety of sectors. Interdisciplinary efforts incorporating knowledge from physics, chemistry, materials science, and engineering are required because of the complex nature of sensing. We may investigate new applications and hasten the creation of innovative sensor designs by promoting a collaborative environment.

References

1. Carrete, J., Li, W., Lindsay, L., Broido, D.A., Gallego, L.J., Mingo, N.: Physically founded phonon dispersions of few-layer materials and the case of borophene. Mater. Res. Lett. **4**, 204–211 (2016). https://doi.org/10.1080/21663831.2016.1174163
2. Ranjan, P., Lee, J.M., Kumar, P., Vinu, A.: Borophene: new sensation in flatland. Adv. Mater. (2020). https://doi.org/10.1002/adma.202000531
3. Hou, C., Tai, G., Liu, Y., Liu, X.: Borophene gas sensor. Nano Res **15**, 2537–2544 (2022). https://doi.org/10.1007/s12274-021-3926-6
4. Khan, K., Tareen, A.K., Aslam, M., et al.: Synthesis, properties and novel electrocatalytic applications of the 2D-borophene Xenes. Prog. Solid State Chem. (2020). https://doi.org/10.1016/j.progsolidstchem.2020.100283
5. Rahman, A., Rahman, M.T., Chowdhury, M.A., Bin Ekram, S., Uddin, M.M.K., Islam, M.R., Dong, L.: Emerging 2D borophene: synthesis, characterization, and sensing applications. Sens Actuators Phys. (2023). https://doi.org/10.1016/j.sna.2023.114468
6. Wang, Z.Q., Lü, T.Y., Wang, H.Q., Feng, Y.P., Zheng, J.C.: Review of borophene and its potential applications. Front Phys. (Beijing) (2019). https://doi.org/10.1007/s11467-019-0884-5
7. Jalali-Mola, Z., Jafari, S.A.: Kinked plasmon dispersion in borophene-borophene and borophene-graphene double layers. Phys. Rev. B (2018). https://doi.org/10.1103/PhysRevB.98.235430
8. Rubab, A., Baig, N., Sher, M., Sohail, M.: Advances in ultrathin borophene materials. Chem. Eng. J. (2020). https://doi.org/10.1016/j.cej.2020.126109
9. Abid, N., Khan, A.M., Shujait, S., Chaudhary, K., Ikram, M., Imran, M., Haider, J., Khan, M., Khan, Q., Maqbool, M.: Synthesis of nanomaterials using various top-down and bottom-up approaches, influencing factors, advantages, and disadvantages: a review. Adv. Colloid Interface Sci. (2022). https://doi.org/10.1016/j.cis.2021.102597

10. Ou, M., Wang, X., Yu, L., Liu, C., Tao, W., Ji, X., Mei, L.: The emergence and evolution of borophene. Adv. Sci. (2021). https://doi.org/10.1002/advs.202001801
11. Chen, C., Lv, H., Zhang, P., et al.: Synthesis of bilayer borophene. Nat. Chem. **14**, 25–31 (2022). https://doi.org/10.1038/s41557-021-00813-z
12. Ranjan, P., Sahu, T.K., Bhushan, R., Yamijala, S.S.R.K.C., Late, D.J., Kumar, P., Vinu, A.: Freestanding borophene and its hybrids. Adv. Mater. (2019). https://doi.org/10.1002/adma.201900353
13. Liu, T., Chen, Y., Zhang, M., Yuan, L., Zhang, C., Wang, J., Fan, J.: A first-principles study of gas molecule adsorption on borophene. AIP Adv. (2017). https://doi.org/10.1063/1.5005959
14. Shukla, V., Wärnå, J., Jena, N.K., Grigoriev, A., Ahuja, R.: Toward the realization of 2D borophene based gas sensor. J. Phys. Chem. C **121**, 26869–26876 (2017). https://doi.org/10.1021/acs.jpcc.7b09552
15. Ge, L., Mu, X., Tian, G., Huang, Q., Ahmed, J., Hu, Z.: Current applications of gas sensor based on 2-D nanomaterial: a mini review. Front. Chem. (2019). https://doi.org/10.3389/fchem.2019.00839
16. Zhao, J., Buldum, A., Han, J., Lu, J.P.: Gas molecule adsorption in carbon nanotubes and nanotube bundles (2002). https://doi.org/10.1088/0957-4484/13/2/312
17. Kadhim, I.H., Hassan, H.A., Ibrahim, F.T.: Hydrogen gas sensing based on nanocrystalline SnO2 thin films operating at low temperatures. Int. J. Hydrogen Energy **45**, 25599–25607 (2020). https://doi.org/10.1016/j.ijhydene.2020.06.136
18. Khan, M.I., Aziz, S.H., Majid, A., Rizwan, M.: Computational study of borophene/boron nitride (B/BN) interface as a promising gas sensor for industrial affiliated gasses. Phys. E Low Dimens. Syst Nanostruct. (2021). https://doi.org/10.1016/j.physe.2021.114692
19. Li, W., Jiang, Q., Li, D., Ao, Z., An, T.: Density functional theory investigation on selective adsorption of VOCs on borophene. Chin. Chem. Lett. **32**, 2803–2806 (2021). https://doi.org/10.1016/j.cclet.2021.01.026
20. Huang, C.S., Murat, A., Babar, V., Montes, E., Schwingenschlögl, U.: Adsorption of the gas molecules NH3, NO, NO2, and CO on borophene. J. Phys. Chem. C **122**, 14665–14670 (2018). https://doi.org/10.1021/acs.jpcc.8b03811
21. Ta, L.T., Hamada, I., Morikawa, Y., Dinh, V.A.: Adsorption of toxic gases on borophene: Surface deformation links to chemisorptions. RSC Adv. **11**, 18279–18287 (2021). https://doi.org/10.1039/D1RA02738G
22. Coropceanu, V., Chen, X.K., Wang, T., Zheng, Z., Brédas, J.L.: Charge-transfer electronic states in organic solar cells. Nat. Rev. Mater. **4**, 689–707 (2019). https://doi.org/10.1038/s41578-019-0137-9
23. Shen, J., Yang, Z., Wang, Y., Xu, L.C., Liu, R., Liu, X.: The gas sensing performance of borophene/MoS2 heterostructure. Appl. Surf. Sci. (2020). https://doi.org/10.1016/j.apsusc.2019.144412
24. Rajkumar, K., Kumar, R.T.R.: Gas sensors based on two-dimensional materials and its mechanisms. In: Fundamentals and Sensing Applications of 2D Materials. Elsevier, pp. 205–258 (2019). https://doi.org/10.1016/B978-0-08-102577-2.00006-3
25. Feng, P.X., Chavez, E., Malca, C.: Super stable pollution gas sensor based on functionalized 2D boron nitride nanosheet materials for high humidity environments. Chemosensors (2018). https://doi.org/10.3390/chemosensors6040049
26. Chaudhary, V., Ashraf, N., Khalid, M., Walvekar, R., Yang, Y., Kaushik, A., Mishra, Y.K.: Emergence of MXene–polymer hybrid nanocomposites as high-performance next-generation chemiresistors for efficient air quality monitoring. Adv. Funct. Mater. (2022). https://doi.org/10.1002/adfm.202112913
27. Weimar, J., Köhli, M., Budach, C., Schmidt, U.: Large-scale boron-lined neutron detection systems as a 3He alternative for cosmic ray neutron sensing. Front. Water (2020). https://doi.org/10.3389/frwa.2020.00016
28. Lv, R., Chen, G., Li, Q., et al.: Ultrasensitive gas detection of large-area boron-doped graphene. Proc. Natl. Acad. Sci. USA **112**, 14527–14532 (2015). https://doi.org/10.1073/pnas.1505993112

29. Kumar Sharma, P., Ruotolo, A., Khan, R., Mishra, Y.K., Kumar Kaushik, N., Kim, N.Y., Kumar Kaushik, A.: Perspectives on 2D-borophene flatland for smart bio-sensing. Mater. Lett. (2022). https://doi.org/10.1016/j.matlet.2021.131089
30. Maier, D., Laubender, E., Basavanna, A., Schumann, S., Güder, F., Urban, G.A., Dincer, C.: Toward continuous monitoring of breath biochemistry: a paper-based wearable sensor for real-time hydrogen peroxide measurement in simulated breath. ACS Sens. **4**, 2945–2951 (2019). https://doi.org/10.1021/acssensors.9b01403
31. Li, J., Zhao, T., Wang, P., Yoon, S., Yu, Y.: Effects of various partitions on the accuracy of virtual in-situ calibration in building energy systems. J. Build. Eng. (2020). https://doi.org/10.1016/j.jobe.2020.101538
32. Manunza, I., Sulis, A., Bonfiglio, A.: Pressure sensing by flexible, organic, field effect transistors. Appl. Phys. Lett. (2006). https://doi.org/10.1063/1.2357924
33. Wang, H., Li, Q., Gao, Y., Miao, F., Zhou, X.F., Wan, X.G.: Strain effects on borophene: ideal strength, negative Possion's ratio and phonon instability. New J. Phys. (2016). https://doi.org/10.1088/1367-2630/18/7/073016
34. Yadav, S., Sadique, M.A., Kaushik, A., Ranjan, P., Khan, R., Srivastava, A.K.: Borophene as an emerging 2D flatland for biomedical applications: Current challenges and future prospects. J. Mater. Chem. B **10**, 1146–1175 (2022). https://doi.org/10.1039/D1TB02277F
35. Chen, S.W., Huang, S.M., Wu, H.S., et al.: A facile, fabric compatible, and flexible borophene nanocomposites for self-powered smart assistive and wound healing applications. Adv. Sci. (2022). https://doi.org/10.1002/advs.202201507
36. Hou, C., Tai, G., Liu, Y., Wu, Z., Liang, X., Liu, X.: Borophene-based materials for energy, sensors and information storage applications. Nano Res. Energy (2023). https://doi.org/10.26599/NRE.2023.9120051
37. Li, D., Gao, J., Cheng, P., He, J., Yin, Y., Hu, Y., Chen, L., Cheng, Y., Zhao, J.: 2D Boron sheets: structure, growth, and electronic and thermal transport properties. Adv. Funct. Mater. (2020). https://doi.org/10.1002/adfm.201904349
38. Qu, X., Yang, R., Tong, F., Zhao, Y., Wang, M.H.: Hierarchical ZnO microstructures decorated with Au nanoparticles for enhanced gas sensing and photocatalytic properties. Powder Technol. **330**, 259–265 (2018). https://doi.org/10.1016/j.powtec.2018.02.019
39. Tu, X., Xu, H., Wang, X., Li, C., Fan, G., Chu, X.: First-principles study of pristine and Li-doped borophene as a candidate to detect and scavenge SO2 gas. Nanotechnology (2021). https://doi.org/10.1088/1361-6528/abfabc
40. Mohn, J., Guggenheim, C., Tuzson, B., Vollmer, M.K., Toyoda, S., Yoshida, N., Emmenegger, L.: A liquid nitrogen-free preconcentration unit for measurements of ambient N2O isotopomers by QCLAS. Atmos Meas. Tech. **3**, 609–618 (2010). https://doi.org/10.5194/amt-3-609-2010
41. Zhang, G., Xie, C., Zhang, S., Zhao, J., Lei, T., Zeng, D.: Temperature-programmed technique accompanied with high-throughput methodology for rapidly searching the optimal operating temperature of MOX gas sensors. ACS Comb. Sci. **16**, 459–465 (2014). https://doi.org/10.1021/co500054r
42. Wu, Z., Tai, G., Liu, R., Hou, C., Shao, W., Liang, X., Wu, Z.: Van der Waals epitaxial growth of borophene on a mica substrate toward a high-performance photodetector. ACS Appl. Mater. Interfaces **13**, 31808–31815 (2021). https://doi.org/10.1021/acsami.1c03146
43. Mannix, A.J., Zhang, Z., Guisinger, N.P., Yakobson, B.I., Hersam, M.C.: Borophene as a prototype for synthetic 2D materials development. Nat. Nanotechnol. **13**, 444–450 (2018). https://doi.org/10.1038/s41565-018-0157-4
44. Hou, C., Tai, G., Liu, B., Wu, Z., Yin, Y.: Borophene-graphene heterostructure: preparation and ultrasensitive humidity sensing. Nano Res. **14**, 2337–2344 (2021). https://doi.org/10.1007/s12274-020-3232-8

Borophene: An Emerging Nanomaterial for Biomedical Applications

Sagar Salave, Nimeet Desai, Siddharth Salve, Dhwani Rana, Shreya Pande, Pranav Patel, Shubhangi Khode, Derajram Benival, and Nagavendra Kommineni

Abstract The emergence of 2D boron nanosheets as a highly promising class of nanomaterials has sparked significant interest in the scientific community. With their exceptional qualities encompassing mechanical strength, thermal conductivity, catalytic activity, anisotropic electrical conductivity, and tunable bandgap, these nanosheets have demonstrated remarkable versatility across a wide range of applications. This chapter provides a thorough and insightful overview of characterization techniques for 2D boron nanosheets including their potential in various fields. This chapter serves as a comprehensive resource, providing valuable insights into their extraordinary qualities and multidisciplinary potential in drug delivery and biomedical applications.

Keywords Boron nanosheets · Synthesis · Drug delivery · Biomedical · Theranostics · Nanotechnology

1 Introduction

The realm of nanotechnology was first introduced by R. Feynman in 1959 in his article "There is Plenty of Room at the Bottom." This ground-breaking work shed light on the immense potential of manipulating matter at the nanoscale, giving birth to a field that holds great promise for innovation. Recent advancements in nanotechnology have highlighted the crucial role played by nanomaterial qualities in the development of future advanced devices. With dimensions ranging from 1 to 100 nm, nanotechnology

S. Salave · N. Desai · S. Salve · D. Rana · P. Patel · S. Khode · D. Benival
National Institute of Pharmaceutical Education and Research (NIPER), Ahmedabad 382355, India

S. Pande
Department of Biomedical Engineering, Indian Institute of Technology Hyderabad, Kandi 502285, India

N. Kommineni (✉)
Center for Biomedical Research, Population Council, New York, NY 10065, USA
e-mail: nagavendra.kommineni@gmail.com

R. Khan et al. (eds.), *2D Boron Nanosheets*, Engineering Materials,
https://doi.org/10.1007/978-981-97-7014-4_8

utilizes atoms and molecules to create materials with unique properties that surpass those of bulk materials. These nanomaterials find applications in various fields, from medicine to industries [1, 2].

To better understand nanomaterials, they can be categorized into four primary types based on their dimensions: zero-dimensional (0D), one-dimensional (1D), and two-dimensional (2D). The properties of these materials are determined by their atomic structure and dimensions [3]. 0D nanomaterials, such as metallic nanoparticles and quantum dots, are fascinating due to their unique physical, optical, chemical, and electrical properties. In addition, 1D nanomaterials like carbon nanotubes and nanofibers are being investigated for various applications [4, 4]. However, the scientific literature on the practical applications of 2D nanomaterials is currently limited compared to 0D and 1D materials.

The discovery of graphene, an atom-thick carbon sheet, by Novoselov, Geim, and co-workers in 2004 revolutionized nanomaterial research [6]. This breakthrough sparked interest in 2D materials with exceptional properties. Researchers have explored various 2D materials, such as derivatives of graphene, black phosphorous, boron, transition metal dichalcogenides, and MXene [7]. These materials offer unique characteristics like tunable electronic structures, large surface areas, mechanical robustness, and easy integration into devices [8]. However, the search for alternative 2D nanomaterials continues to address graphene's limitations, such as its lack of a bandgap and non-biodegradability. In a recent high-throughput computational study, it was reported that there are 5,619 monolayer compounds, out of which 1,825 can be easily exfoliated. A subset of 56 compounds, featuring half-metals and half-semiconductors, was further investigated for their electronic, magnetic, topological, and vibrational properties [9].

2D nanomaterials consist of one (or two) atom-thick layers, offering a higher surface-to-volume ratio. The distinctive properties of these nanomaterials stem from various factors, including their specific shape, high surface energy, and surface charge. The structural characteristics profoundly influence their overall behavior and present intriguing opportunities for tailored functionalities in a wide range of applications. Most 2D nanostructures are composed of multiple layers held together by weak van der Waals forces. The compact structure of these materials provides excellent stability. By employing physical and chemical methods, layer stacking can be disrupted, leading to delamination or exfoliation. Delamination increases interlayer spacing, allowing the infiltration of other nanomaterials or biomolecules. Exfoliation rearranges the layers, which is crucial for creating nanohybrids or nanocomposites [10]. 2D nanomaterials exhibit different crystalline phases, arrangements, and structures, earning them names such as nanosheets, nanodisks, nano prisms, and nanorings. These materials also hold promise for advanced energy storage systems, enabling high-performance batteries and supercapacitors with enhanced energy density and faster charging capabilities. This unique arrangement of atoms imparts mechanical flexibility, optical transparency, and electronic conductivity [11].

Graphene has found applications in therapeutics, diagnostics, sensors, and energy storage, but recently, research attention has shifted towards 2D boron and other non-carbon-based nanosheets. These nanosheets exhibit exceptional properties approaching the limits of known materials, driven by their ultrathin structure and quantum confinement effects. Ongoing efforts in synthesizing graphene-analogous nanosheets increasingly focus on 2D boron due to its growing scientific community interest.

Boron is an element of group III of the periodic table, situated between metallic beryllium and non-metallic carbon. It has the electronic configuration of $1s^2 2s^2 2p^1$. The $2s^2$ electrons of the boron impart non-metallic characteristics, whereas the $2p^1$ electron favors metallic properties [12]. This unique configuration of boron allows it to form covalent two-center bonds, as well as metallic-like multi-center bonds, resulting in the formation of various nanostructures, viz., zero-dimensional (0D) nanoclusters, one-dimensional (1D) nanotubes, and 2D nanosheets. The boron structure formed as a result of the arrangement of a few atomic clusters shows a planar structure (benzene-ring-like) whereas the films and tubes are formed as a result of the arrangement of slightly larger boron atomic clusters [13]. The 2D boron structures mainly include boron nanosheets and structures of boron that are single atom thick, known as borophene [14]. Boron nitride, also known as 'white graphene' shows similar structural, physicochemical, and functional properties as compared to boron nanosheets. While boron nanosheets consist solely of boron atoms, boron nitride nanosheets are composed of a hexagonal lattice of alternating boron and nitrogen atoms. This elemental variation leads to stark differences in their properties and potential applications. Boron nanosheets exhibit unique electronic and mechanical properties, whereas boron nitride nanosheets are known for their exceptional thermal stability, high electrical insulation, and chemical inertness [15–17].

The unique properties of 2D boron nanosheets make them versatile in different fields of nanotechnology and applications. Due to their unique electronic structure, they possess tunable bandgaps, which can be advantageous for developing electronic devices, sensors, photodetectors, and optoelectronic applications [18–20]. Unlike the perfect hexagonal lattice structure of graphene, boron nanosheets exhibit more complex arrangements of polygons, including triangles, hexagons, and heptagons. These diverse ring structures introduce curvature to the nanosheets, influencing their electronic band structure and mechanical properties. The presence of defects and edges in the boron lattice further enhances their catalytic properties, making them potential candidates for heterogeneous catalysis applications [21, 22]. The high thermal conductivity and mechanical strength of the 2D boron nanosheets open up possibilities for thermal management, nanoelectromechanical systems (NEMS), and reinforcing composite materials [23, 24]. The structural features of boron nanosheets allow them to hold the adsorbed metal and facilitate hydrogen storage. Their low-dimensional nature, large surface area, and low mass densities make them suitable to act as electrodes, can be employed as supercapacitors, and can also be used in lithium-ion batteries [25–27].

As the field of 2D boron nanosheets is relatively nascent, there are several challenges and opportunities to explore. The synthesis methods, scalability, stability, and functionalization strategies are among the key areas of investigation. Addressing these challenges will not only advance the fundamental understanding of boron nanosheets but also pave the way for their widespread applications in various industries. In this chapter, we have discussed the characterization techniques employed for obtaining 2D boron nanosheets, explored their unique properties, and their potential applications across different fields. Further, drug delivery and biomedical applications have been discussed thoroughly.

2 Characterization Techniques for 2D Boron Nanosheets

For further modification of the structures, it is crucial to comprehend the physical and chemical properties of 2D boron nanosheets. The most popular and efficient microscopic and spectroscopic methods for measuring these materials' size, shape, number of layers, chemical composition, chemical bonding, and other properties have been compiled in the following section.

2.1 Scanning Probe Microscopy Techniques

AFM is a very useful tool that uses the attraction and repulsive interactions between atoms to study the surfaces of materials and measure the thickness of nanosheets. Unlike Scanning Tunneling Microscopy (STM), which depends on the sample being examined, AFM offers direct topographical information without confusing the electrical structure [28]. AFM scans a sample's surface with a pointed probe and measures the forces exerted by the probe against the material. AFM can generate high-resolution pictures of the sample surface by measuring the forces [29]. AFM can be utilized in the instance of 2D boron nanosheets to determine the morphology and thickness of the nanosheets [30]. Without requiring considerable sample preparation because AFM may function in air or liquid conditions, it is possible to photograph the nanosheets in their natural state [31].

2.2 UV–Visible Spectroscopy

A useful method for describing 2D boron nanosheets is UV–visible spectroscopy. UV–visible spectroscopy may reveal details on the bandgap of boron nanosheets, which is a crucial factor in determining their electrical and optical properties [32]. The energy needed to excite an electron from the valence band to the conduction band is known as the bandgap, and it affects a material's capacity to both absorb and emit light. A bandgap estimate for boron nanosheets can be made using the UV–visible absorption spectrum [33]. The amount of energy needed to excite an electron from the valence band to the conduction band, or bandgap, controls how much light a material can absorb and emit. The absorption spectra obtained from UV–visible spectroscopy can be used to measure the bandgap of boron nanosheets [33]. Therefore, by utilizing UV–visible spectroscopy, one can effectively determine the bandgap, size, and thickness of 2D boron nanosheets, which are crucial parameters influencing their electrical and optical properties.

2.3 *Fourier Transform Infrared (FTIR) Spectroscopy*

FTIR is a useful technique for the characterization of 2D boron nanosheets. FTIR helps measure a substance's infrared radiation transmission or absorption as a function of wavelength or frequency. In addition to providing details on the crystal structure and orientation, this information can be utilized to detect functional groups and chemical bonds in a material [34]. The stability and electrical characteristics of 2D boron nanosheets depend on the presence of boron-oxygen (B-O) and boron–hydrogen (B–H) bonds, which may be detected using FTIR [28].

2.4 *Raman Spectroscopy*

Raman spectroscopy is a common and powerful technique to analyze 2D materials and has been widely used to characterize bulk boron polymorphs. The complex chemical and structural characteristics of various boron polymorphs can be distinguished by their distinctive Raman spectra, which have been frequently utilized to describe bulk boron polymorphs (α-, β-, and γ-B). Additionally, the line width of the Raman modes can be used to estimate the EPC strength, which can then be used to compute Tc using the McMillan equation. This method can provide data on Debye temperature and thermal conductance, which are crucial for comprehending the thermodynamic characteristics of borophene.

2.5 *X-Ray Diffraction Analysis (XRD)*

The diffraction pattern of the nanosheets is obtained using XRD. For XRD investigation, the 2D boron nanosheets must be evenly distributed on a substrate and carefully positioned [35]. The sample is exposed to X-rays, and the pattern that emerges reveals details about the nanosheets' orientation and crystal structure. The crystal structure, orientation, and lattice parameters of the nanosheets can be ascertained by analyzing the diffraction pattern produced by XRD examination. In this, the observed diffraction pattern is contrasted with simulated patterns produced from well-known crystal structures. To ascertain the preferred orientation of the nanosheets, texture analysis can be used on the diffraction pattern. This reveals details about the material's crystalline anisotropy.

2.6 Scanning Electron Microscope (SEM) Analysis

A concentrated stream of high-energy electrons is utilized by the SEM to produce a variety of signals at the surface of solid specimens. In addition to the sample's exterior morphology (texture), chemical composition, crystalline structure, and orientation of its constituent materials, the signals resulting from electron-sample interactions also provide information about the sample.

2.7 Energy Dispersive X-Ray Spectroscopy (EDX)

EDX is a method that can be used to ascertain a material's elemental composition. The presence and distribution of boron and other elements within the material can be determined in the instance of 2D boron nanosheets using EDX [36]. When a material is hit with electrons, it emits distinctive X-rays, which are what the EDX device detects. The elements that are present in the material and their relative concentrations are then determined using the energy and intensity of these X-rays [37].

3 Applications of 2D Boron Nanosheets

In recent years, 2D materials have gained a lot of attention due to their distinct characteristics and potential uses in a variety of sectors. One interesting option for a variety of applications among these 2D materials is the 2D boron nanosheet. 2D boron nanosheets have been investigated for applications in electronics, energy storage, catalysis, sensing, biomedicine, and other fields due to their remarkable electrical, mechanical, thermal, and optical properties.

3.1 Electronics and Optoelectronic Applications

Recent years have seen a lot of research activity in the production and characterization of boron quantum dots (BQDs) embedded in 2D boron nanosheets. Due to the quantum confinement phenomenon, boron quantum dots, which are tiny clusters of boron atoms, have optical and electrical capabilities. These BQDs have several benefits over conventional quantum dots when they are incorporated into

2D boron nanosheets [38]. Moreover, borophene can be used for flexible electronic and bendable devices such as flexible displays, wearable devices, and electronic textiles due to its mechanical as well as electronic properties. The electronic component is made of borophenes that conform to curves and irregular surfaces due to their high conductivity and flexibility [5].

3.2 Energy Application

High capacity, adequate voltage, and good electronic and ionic conductivity are essential criteria for electrode materials. 2D boron nanosheets meet these qualities and are widely used for batteries and capacitors. 2D boron nanosheets are a lightweight material and have high surface activity. Due to these properties, it is advantageous to achieve an ultrahigh capacity of Li/Na/Mg/AI storage and is the best candidate for Li/Na/K/Mg/AI ion batteries. For Li-ion batteries, 2-Pmmn borophane is an excellent anode material.

3.3 Nanoelectromechanical System (NEMS)

The excellent mechanical properties and electrical conductivity of boron nanosheets employed as strain sensors, pressure sensors, or gas sensors can be used to detect minute mechanical and environmental changes by measuring the corresponding electrical signal. The unique mechanical properties of boron nanosheets, such as high flexibility, strength, and tunable electronic properties, make them a good candidate for actuation at the nanoscale, such as positioning, switching, or pumping. Boron nanosheets can undergo controlled deformation, enabling precise actuation by applying electrical and mechanical stimuli. Boron nanosheets exhibit piezoelectric characteristics, making them capable of transforming mechanical energy into electrical energy, which is valuable for energy harvesting applications. Leveraging their piezoelectric properties, these nanosheets can be seamlessly integrated into NEMS devices to effectively capture and convert environmental mechanical energy into electrical energy.

3.4 Catalysis Application

2D boron nanosheets are investigated as potential catalysts due to their high surface area, tunable electronic properties, and high catalytic properties. Owing to their distinct electrical characteristics and large surface area, 2D boron nanosheets have demonstrated excellent hydrogen evolution reaction catalytic activity. They are thus promising hydrogen fuel cell catalysts [39]. Additionally, it has been demonstrated that boron nanosheets are efficient catalysts for the oxygen reduction reaction, a critical process in fuel cells and other electrochemical systems [40].

3.5 Coating Application

Due to the excellent mechanical strength and chemical stability of boron nanosheets, they can be incorporated into coatings to provide enhanced corrosion resistance. It can improve the longevity and performance of metals, alloys, and other susceptible materials by coating the surface [41]. 2D boron nanosheets can be used for thermal protection due to their better thermal conductivity and stability at high temperatures. These coatings can help regulate temperature, protect the surface from thermal damage, and enhance the thermal management of the device; therefore, they are widely used where heat dissipation or insulation is required [42].

4 Drug Delivery and Biomedical Applications of 2D Boron Nanosheets

The following section summarises the various biomedical and drug-delivery applications of borophene (Fig. 1).

4.1 Drug Carrier Systems

Several approaches have been reported as drug delivery carrier systems [2, 5, 43–68]. Recently, borophene has emerged as a promising and innovative drug delivery platform, as underscored by the insightful investigations presented in the work by Vatanparast et al. Utilizing a comprehensive approach involving Density Functional Theory (DFT) calculations and Molecular Dynamics (MD) simulations, the study delves into the fundamental interactions between three distinct anticancer drugs, namely 5-fluorouracil (FU), 6-mercaptopurine (MP), and 6-thioguanine (TG), with hexagonal boron nitride nanosheets (BNNSs). Notably, the study unveils that the adsorption process of these drugs onto BNNSs is both spontaneous and exothermic, indicative of a favorable affinity. Furthermore, intriguing variations in binding energy magnitudes between the drugs are disclosed, with TG and MP demonstrating superior affinities for the BNNS surface compared to FU. This variation is coupled with an augmented polarity within drug-loaded complexes, signifying a potential avenue for enhancing solubility—a critical factor governing effective drug delivery within biological milieus. Quantum molecular descriptor analyses divulge heightened chemical reactivity upon drug adsorption, augmenting therapeutic prospects.

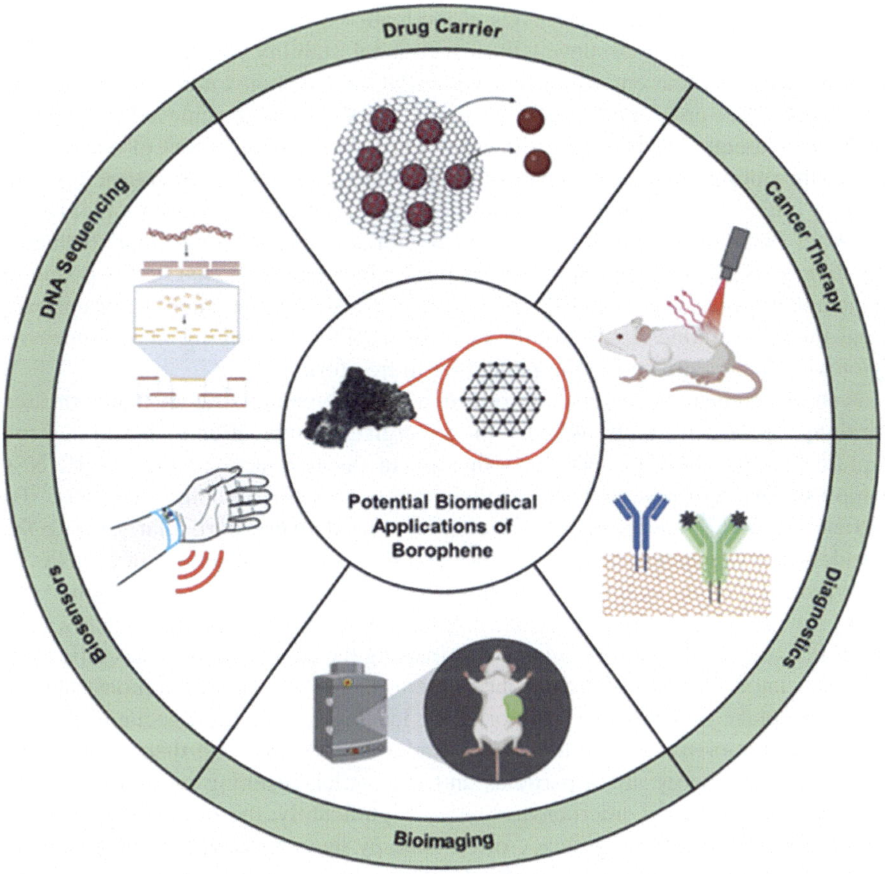

Fig. 1 Potential drug delivery and biomedical applications of borophene

An incisive examination of intermolecular interactions through noncovalent interaction analyses elucidates the intricate forces governing drug-nanosheet bonding. Equally noteworthy is the discernment that dispersion interactions predominantly underpin the stability of drug-BNNS complexes. MD simulations underscore the distinct interaction energies across different drug-nanosheet systems, suggesting the prominence of the BN-TG interface in terms of interactions. Notably, in simulated acidic conditions, the interaction energies tend to diminish, implying a plausible mechanism for targeted drug release within cancer cells [69].

In a similar study, Duverger et al. employed a rigorous combination of experimental data, DFT, time-dependent density functional theory (TDDFT), and MD simulations to elucidate the intricate behavior of Boron nitride oxide nanosheets (BNO-NS) as a vehicle for the anticancer drug doxorubicin (DOX). Notably, the research establishes that non-covalent interactions between DOX and BNO-NS preserve DOX's essential UV–Vis response and chemisorption properties, ensuring that the photodynamic characteristics of the loaded drug remain unaffected. Moreover, the MD simulations unveil the exceptional stability of the DOX/BNO-NS system across diverse environments, with BNO-NS demonstrating its efficacy in stabilizing DOX proximate to cellular membranes—a feat unachieved by isolated DOX. This unique attribute is attributed to the strategic deployment of chemisorption as the optimal binding mechanism, affirming its pivotal role in enhancing drug delivery efficacy. As an innovative stride, the study pioneers the exploration of DOX/BNO-NS behavior in proximity to cell membranes through molecular dynamics simulations. This contribution augments our understanding of BNO-NS's potential in therapeutic agent delivery [70]. The same group also reported trapping the antibacterial fluoroquinolone agent ciprofloxacin on BNO-NS, resulting in an improved cytotoxic effect compared to the same drug in free form [71].

A study conducted by Li et al. introduces a groundbreaking strategy that harnesses the unique properties of BNNSs for the synergistic combination of boron neutron capture therapy (BNCT) and chemotherapy in the treatment of tumors. BNNSs exhibit exceptional potential due to their ability to serve as both boron-10 (^{10}B) carriers for BNCT and effective vehicles for drug delivery, particularly for DOX, a widely used chemotherapeutic agent. The dual-functional nature of BNNSs is underpinned by their intrinsic properties, enabling precise drug release upon exposure to neutron irradiation. This neutron-triggered release mechanism allows for controlled and localized drug delivery, addressing the challenge of nonspecific toxicity associated with conventional chemotherapy. In vitro studies demonstrate the effective uptake of BNNSs by tumor cells, leading to enhanced accumulation of ^{10}B within tumor tissues. Furthermore, irradiating BNNSs with thermal neutrons generates high-energy alpha particles and ^{7}Li nuclei, resulting in significant cell death within the tumor microenvironment. Significantly, the co-loading of DOX onto BNNSs enhances the efficacy of the therapy by capitalizing on the enhanced permeability and retention (EPR) effect. Neutron irradiation facilitates the release of DOX from the BNNS carrier, leading to dual-action treatment effects. In vivo experiments conducted in a murine triple-negative breast cancer model showcase the potential of this strategy. The results reveal remarkable suppression of tumor growth after combined radiochemotherapy employing DOX-loaded BNNSs and neutron irradiation (Fig. 2a). This combinational treatment approach outperforms monotherapies and achieves a notable therapeutic outcome. Histological analysis corroborates the effectiveness of the treatment, as evidenced by substantial tumor damage while preserving the typical morphology of major organs [72].

In a recent study, a salt-template method was employed by Cheng et al. to synthesize gram-scale quantities of pure and crystalline BNNSs, yielding quantities surpassing those achievable through conventional lab-based techniques by orders of magnitude. These BNNSs, once chemically modified with hydroxyl groups, exhibited remarkable attributes conducive to drug delivery systems. Specifically, the hydroxylated BNNSs (OH-BNNSs) demonstrated exceptional dispersibility and stability in water, remaining free from precipitation for over two days at a concentration of 1 mg mL^{-1}. The biocompatibility of OH-BNNSs was underscored by

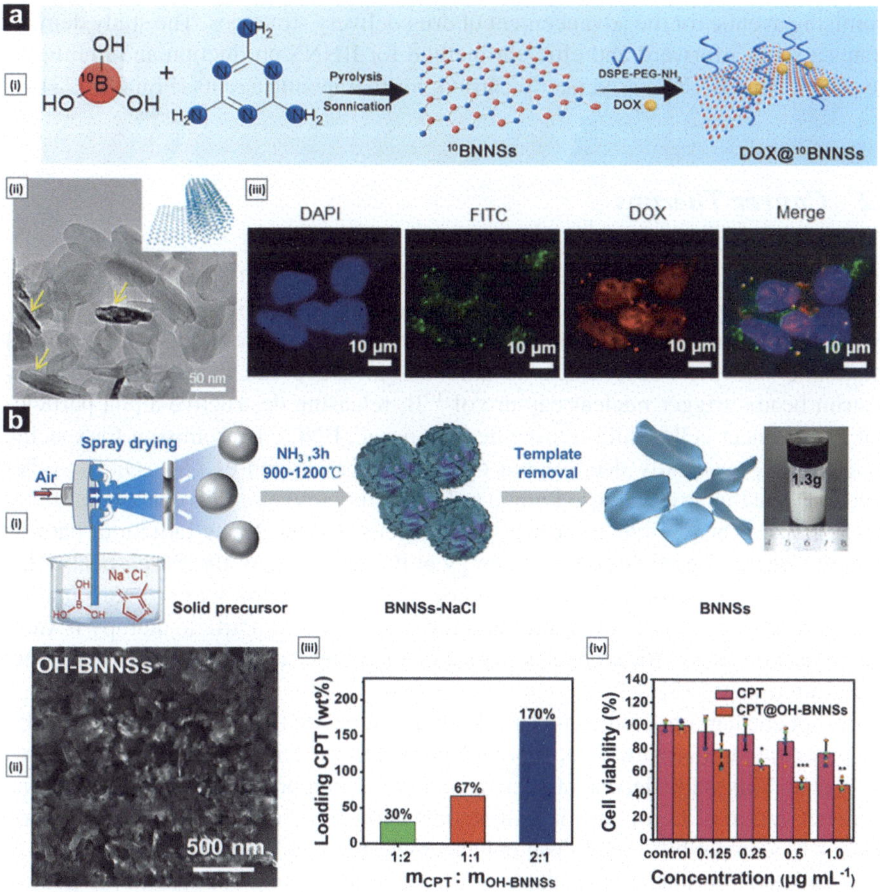

Fig. 2 **a** (i) Schematic depicting the fabrication of ultrathin 2D BNNSs loaded with DOX. (ii) Representative TEM image. (iii) Representative confocal images of 4T1 cells incubated with DOX-loaded BNNSs followed by neutron irradiation. Adapted with permission from [72], copyright Elsevier 2021. **b** (i) Illustrative scheme of salt-template synthesis procedure of BNNSs. (ii) SEM image of OH-BNNSs after treating BNNSs-900 in nitric acid. (iii) CPT loading capacities of the OH-BNNSs at different CPT: OH-BNNS ratios. (iv) Cell viability of free CPT and CPT@OH-BNNSs solutions for 24 h at different concentrations (cell lines used is 4T1). Adapted with permission from [73], Copyright Elsevier 2022

their ability to maintain 99% cell viability when incubated with a sample concentration of 100 μg mL^{-1} for 24 h, affirming their suitability for biomedical applications. Notably, the OH-BNNSs showcased a compelling capacity to load camptothecin (CPT), an anticancer drug, with an impressive loading efficiency of up to 170 wt%. This loading capability paved the way for the effective delivery of CPT, as evidenced by the CPT@OH-BNNSs' superior performance in inhibiting 4T1 breast cancer cells compared to free drugs, both in vitro and in vivo. These findings offer a promising avenue for the advancement of drug delivery strategies. The study demonstrates a straightforward and efficient method for BNNS production and highlights the potential of BNNSs as versatile carriers for therapeutic agents (Fig. 2b) [73].

4.2 Cancer Therapy

Boron neutron capture therapy (BNCT) is a targeted cancer treatment that exploits ^{10}B nuclei's properties. When ^{10}B absorbs neutrons, it emits potent alpha particles, damaging nearby cancer cells effectively. BNCT involves administering a ^{10}B compound into the body, which accumulates in active cancer cells. Low-energy neutron beams trigger nuclear capture of ^{10}B, releasing destructive alpha particles that harm cancer cells while sparing healthy tissue. BNCT is promising for specific cancer types, especially deep-seated or near-critical structures [74]. On the other hand, particle therapy (PT, also known as Hadrontherapy) employs charged particles like protons or heavy ions to destroy cancer cells. Unlike X-ray radiation, particle therapy's charged particles deposit energy at the target's end, the Bragg peak. This enables precise dose delivery to the tumor while minimizing harm to surrounding tissues, resulting in better outcomes and fewer side effects. Particle therapy is valuable for tumors near critical areas or in pediatric patients, reducing radiation exposure to healthy tissues [75].

Borophene has attracted considerable attention within the realm of BNCT and PT as a potential radiosensitizer, as expounded upon in a recent investigation conducted by Qi et al. This study delves into pertinent aspects, comprising (1) the synthesis and characterization of borophene nanomaterials and (2) the comprehensive evaluation of their biocompatibility and dose enhancement potential. Employing a distinctive methodology, borophene is synthesized through ball-milling-assisted sonochemical liquid-phase exfoliation, effectively circumventing the limitations inherent in vapor-based techniques. The synthesis procedure involves the dispersion of bulk boron powder within a ball mill, culminating in the generation of layered boron sheets following a sequence of meticulously executed centrifugation steps. The synthesized borophene exhibits a discernibly crystalline nature, as validated by high-resolution transmission electron microscopy (HRTEM) data, which reveals a distinct d-spacing of 0.25 nm corresponding to the X3 phase of the boron structure. Furthermore, the investigation methodically examines borophene's biocompatibility through a series of cytotoxicity assays, showcasing survival rates exceeding 80% even at an elevated concentration of 250 $\mu g\ ml^{-1}$. Within the purview of PT and BNCT scenarios, the study meticulously elucidates borophene's role as a robust radiosensitizer. This is achieved through the meticulous employment of Monte-Carlo simulations, which remarkably elucidate its efficacy in enhancing doses. Notably, the dose enhancement ratios (DER) consistently surpass 2 in BNCT, even when considering relatively low boron concentrations of 100 $\mu g\ g^{-1}$. The fundamental basis for this amplified dose effect lies in borophene's intrinsic capability to generate alpha particles upon interaction with secondary neutrons, thereby substantially amplifying the therapeutic efficacy of the treatment regimen. The study's integration of confocal imaging methodologies further augments its analytical depth by comprehensively evaluating borophene's spatial distribution. This critical step lays the groundwork for insightful microdosimetry investigations. The study also considers the potential implications of utilizing borophene in clinical settings. A central focus is directed toward striking a delicate balance between achieving dose enhancement and managing ROS-mediated cell damage [76].

Cancer photothermal therapy (PTT) and photodynamic therapy (PDT) are advanced techniques that utilize light for precise and selective cancer treatment. PTT employs light-absorbing nanoparticles as photothermal agents. These agents are administered into the body and accumulate within tumors. When exposed to specific light wavelengths, they convert light energy into heat, raising the tumor's temperature and causing localized destruction of cancer cells. This approach offers minimal invasiveness, accurate tumor targeting, and potential synergies with other treatments [77]. PDT involves light-activated compounds called photosensitizers. These substances accumulate in tumors and produce reactive oxygen species that induce cancer cell death upon exposure to specific light wavelengths. The therapy targets tumor cells and disrupts nearby blood vessels, reducing tumor blood supply. PDT's advantages include precise targeting, limited impact on healthy tissue, and potential for repeat treatments [78].

The amalgamation of PTT and PDT for augmented cancer treatment has garnered substantial scientific interest. Particularly noteworthy is the work conducted by Kang et al., which underscores the utility of boron-based nanosheets as a promising platform in this context. These nanosheets, synthesized using an advanced liquid exfoliation technique, exhibit a distinctive two-dimensional architecture characterized by ultrathin layers measuring approximately 2.4 nm thick. This unique structural arrangement facilitates efficient photosensitizer chlorin e6 (Ce6) encapsulation through hydrophobic interactions. To enhance the stability and biocompatibility of these nanosheets, a judicious modification approach is employed, involving the integration of poly(allylamine hydrochloride) (PAH) and poly(acrylic acid) (PAA). The resulting nanostructure, denoted as B@Ce6–PAH–PAA, demonstrates a range of favorable attributes. These encompass dual-mode imaging-guided precision, proficient photothermal conversion capabilities, and adept photodynamic performance. The strategic dimensions of these nanosheets confer an optimal nanoparticulate size, thereby facilitating preferential accumulation within neoplastic tissues. This behavior is attributed to the established EPR effect. Rigorous assessment, encompassing both in vitro and in vivo studies, substantiates the nanosheets' commendable biocompatibility, intracellular uptake kinetics, and remarkable therapeutic efficacy when employed with PTT and PDT modalities. Notably, the concurrent implementation of these therapies culminates in a synergistic therapeutic outcome that surpasses the efficacy of individual interventions. This work underscores the promising potential of boron-based nanosheets as a robust platform for advancing the field of cancer treatment through the integration of PTT and PDT [79].

In a similar study, Kang et al. detailed the creation of boron nanosheets (BNSs) for synergistic low-temperature PTT and chemotherapy in cancer treatment. These top-down synthesized nanosheets were characterized via TEM and AFM with irregular shapes and a single-layer thickness of approximately 6.1 nm. Boron oxide presence from thermal oxidation etching was confirmed using XPS, with distinct B-B and B-O bonds in the B 1s region. Electrostatic adsorption led to incorporating H_2N-PEG-NH_2 groups, forming PEGylated BNSs (B-PEG), subsequently functionalized with cRGD for increased biocompatibility. The integration of B-PEG-cRGD was meticulously described involving an amine-carboxyl reaction. UV–vis and zeta potential measurements validated DOX and 17AAG loading onto B-PEG-cRGD nanosheets via physical adsorption, impacting zeta potential. Photothermal properties were scrutinized under NIR laser, revealing concentration-dependent temperature elevation and consistent attributes over cycles, with calculated PTCE showcasing efficacy. pH and NIR-responsive DOX and 17AAG release from B-PEG-cRGD were explored. Cellular uptake of DOX-17AAG@B-PEG-cRGD was verified via microscopy and flow cytometry, reinforcing cRGD targeting. In vitro, assays highlighted nanomaterial biocompatibility and significant cytotoxicity from DOX-17AAG@B-PEG-cRGD due to 17AAG-mediated Hsp90 inhibition under low-temperature hyperthermia. Concentration-dependent cell growth modulation by 17AAG indicated synergistic PTT/chemotherapy efficacy. In vivo, DOX-17AAG@B-PEG-cRGD showed tumor accumulation, improved therapeutic effects, prolonged survival, and minimal toxicity. Histology and metabolic studies underscored the nano platform's biocompatibility and targeted therapeutic potential [80].

In the domain of cancer immunotherapy, personalized therapeutic vaccines have emerged as a propitious avenue, particularly for immune desert tumors that present considerable challenges due to their constrained neoantigen recognition and immunosuppressive microenvironment. To address these challenges, an innovative strategy has been introduced by Sun et al., capitalizing on boron nanosheet-based nano vaccines (designated as B@TA-R848) to enhance photo-immunotherapy against immune desert tumors, with specific emphasis on triple-negative breast cancer (TNBC). The B@TA-R848 formulation incorporates tumor autoantigens (TA) derived from excised tumors, immobilized onto two-dimensional boron nanosheets via a polydopamine coating. This personalized nano vaccine boasts a multitude of advantageous attributes, encompassing efficient drug delivery and release, capabilities for photoacoustic imaging, and biocompatibility. A notable feature of B@TA-R848 is its heightened photothermal conversion efficiency when exposed to 808 nm laser irradiation, thereby augmenting its therapeutic potential. Leveraging its inherent multifunctionality, this nano vaccine instigates a comprehensive antitumor immune response on a systemic level, effectuates a reshaping of the tumor microenvironment, and bolsters the infiltration of immune cells within the tumor. Crucially, preclinical assessments conducted on a murine model of TNBC showcase the efficacy of B@TA-R848-mediated photo-immunotherapy, culminating in substantial inhibition of tumor growth, recurrence mitigation, and metastasis suppression. Despite the intricacies associated with the immunological complexity of tumors and the heterogeneity among individuals, this study underscores the astute fusion of personalized tumor autoantigens with advanced nanomaterials. This exemplifies the substantial potential of B@TA-R848 nano vaccines in ushering a transformative shift within the therapeutic panorama concerning immune desert tumors. Further investigations into synergistic combination therapies and tailored functional modifications aim to amplify clinical anticancer outcomes and broaden personalized treatments' horizons [81].

An investigation conducted by Xiao et al. makes a significant contribution to the field of ultrasound-mediated synergistic cancer therapy by utilizing boron-based nanosheets (BOP NSs). The authors respond to the imperative demand for enhanced treatment methodologies by meticulously synthesizing and characterizing BOP NSs via a methodical liquid-phase exfoliation procedure. The nanosheets exhibit distinctive characteristics, including heightened sensitivity to ultrasound, increased production of ROS, and precisely regulated biodegradability. This multifaceted approach is thoroughly substantiated by an exhaustive array of analytical techniques, encompassing TEM, AFM, XRD, energy dispersive X-ray spectroscopy, X-ray photoelectron spectroscopy, and TGA, all of which unequivocally confirm the successful synthesis and structural integrity of the nanosheets. The research systematically probes into the degradation dynamics of BOP NSs, elucidating their disintegration upon exposure to oxygen and highlighting their potential for safe and efficient in vivo clearance. Furthermore, the study conducts a comprehensive exploration of the sonodynamic performance exhibited by BOP NSs, underscored by their proficient ROS generation capacity as validated through UV–visible absorption spectra, Singlet Oxygen Sensor Green (SOSG) fluorescence studies, and Electron Spin Resonance (ESR) analyses. The authors extend their inquiry to the peroxidase-like activity of BOP NSs, showcasing their prowess in catalyzing the oxidation of peroxidase substrate TMB and their augmented efficacy in the presence of ultrasound stimulation. Noteworthy achievements include the comprehensive in vitro and in vivo assessments of BOP NSs, unveiling their adeptness in inducing apoptotic and necrotic processes in tumor cells and curbing tumor growth with ultrasound intervention (Fig. 3). These results impeccably exemplify the potential of BOP NSs as an efficacious platform for synergistic cancer therapy [82].

4.3 *Bioimaging and Theranostics*

Bioimaging agents enhance our ability to visualize cancer at various stages, aiding in diagnosis, treatment planning, and monitoring. Theranostics agents, on the other hand, bridge the gap between diagnostics and therapy, enabling personalized and targeted treatment approaches. Together, these agents significantly contribute to advancing cancer care, improving patient outcomes, and reducing the burden of the disease [83].

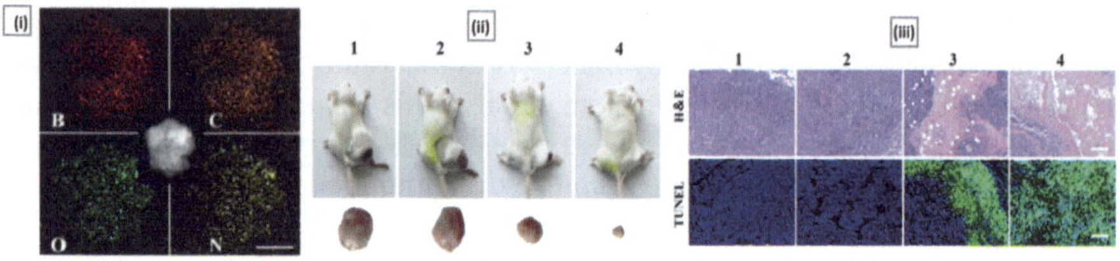

Fig. 3 (i) HAADF-STEM and elemental mapping of B, C, N, and O elements of BOP NSs (Scale bar: 50 nm). (ii) Photograph of mice and tumors and (iii) H&E staining/TUNEL staining of corresponding tumor tissue sections from different treatment groups. [Here 1: PBS, 2: only ultrasound, (3) B-OH@Cy5-PEG-NH_2 nanosheets, (4) B-OH@Cy5-PEG-NH_2 nanosheets + ultrasound] (Scale bar: 100 μm). Adapted with permission from [82], Copyright Elsevier 2022

Ultrathin boron nanosheets have been identified as a compelling photoluminescent material of significant relevance to bioimaging applications, as evidenced by the comprehensive inquiry presented by Ma et al. The investigation highlights a novel technique for liquid-phase exfoliation that combines probe sonication and subsequent water bath sonication in the synthesis of ultrathin boron nanosheets from bulk boron powder. The efficacy of this approach is rooted in the anisotropic processing effects induced by probe sonication, yielding oriented bubbles and selective cleavage of the boron bulk into 2D slices along the vertical axis. The resultant ultrathin boron nanosheets exhibit a consistent size and thickness distribution. The structural integrity and development of surface defects are systematically examined using XRD patterns, high-resolution TEM images, X-ray photoelectron spectroscopy, and Fourier transform infrared spectroscopy. The optical characteristics of these ultrathin boron nanosheets constitute a pivotal focus of the investigation, revealing distinctive blue luminescence emission upon excitation with 365 nm light. The fluorescence emission spectroscopy probes the emission peaks believed to originate from the synergistic interplay of quantum confinement effects and surface defects. Notably, the study elucidates the concentration-dependent behavior of fluorescence emission, identifying an optimal concentration that yields the highest emission intensity. Remarkably, the ultrathin boron nanosheets exhibit outstanding photostability, surpassing conventional dyes, thus establishing their potential as robust imaging agents. This utility is further underscored by femtosecond transient absorption (TA) spectroscopy, which uncovers efficient electron–hole pair separation, resulting in notable photoinduced fluorescence emission. Additionally, the biocompatibility and toxicity profile of the ultrathin boron nanosheets is notably impressive, as confirmed through cell viability assays conducted with cancer cell lines. Successful implementation of ultrathin boron nanosheets for cellular imaging is convincingly demonstrated, as evidenced by bright blue light emission upon excitation at 405 nm [84].

In their pioneering contribution, Ji et al. introduce an innovative top-down synthetic strategy that synergistically integrates thermal oxidation etching and liquid exfoliation techniques, creating ultrathin boron nanosheets with precisely controlled nanoscale thickness. This novel approach involves controlled oxidation of B–B units within bulky boron layers at elevated temperatures, followed by meticulously orchestrated liquid exfoliation stages, culminating in ultrathin B NSs with a planar dimension of 100 nm and an impressive thickness of 5 nm. Subsequent functionalization of the boron nanosheets with amine-functionalized polyethylene glycol ($PEG-NH_2$) enhances biocompatibility and dispersibility, establishing the PEGylated boron nanosheets as a versatile platform at the forefront of cancer therapy and imaging. A notable achievement of this work is the remarkable photothermal conversion efficiency of 42.5% demonstrated by the engineered B-PEG NSs, rendering them suitable for PTT and photothermal/photoacoustic imaging (PAI) applications. Furthermore, these nanosheets exhibit remarkable loading capacities for chemotherapeutic agents, such as DOX, and imaging agents, like cyanine 5.5. Comprehensive experiments were conducted using murine models to assess the potential of B-PEG NSs as a photoacoustic contrast agent. Upon NIR irradiation, a substantial enhancement in photoacoustic signals was observed with increasing B-PEG NSs concentration, validating their proficiency as a photoacoustic reagent. Notably, a robust linear correlation between photoacoustic signals and B-PEG NSs concentration was established, affirming the reliability and promise of these nanosheets for photoacoustic imaging. The consistent amplification of photoacoustic signals over time further demonstrated the enduring and dependable performance of B-PEG NSs in this capacity. Furthermore, the potential of B-PEG NSs for photothermal therapy was evaluated by monitoring the body temperature of mice during NIR irradiation. Encouragingly, mice treated with B-PEG NSs displayed significantly lower body temperatures than untreated controls, confirming the efficient photothermal conversion capability of the nanosheets. A comprehensive assessment of therapeutic outcomes revealed exceptional tumor growth inhibition in mice subjected to a combined chemotherapy-photothermal regimen. Visual comparison of tumor volumes highlighted the pronounced therapeutic efficacy of B-PEG NSs in impeding tumor progression [85].

An investigation conducted by Xie et al. represents an innovative foray into light-triggered tumor theranostics using chemically-exfoliated borophene. The distinct optical, electronic, and chemical properties of borophene, with its varied polymorphic configurations, including a notable 2D metallic phase, form the basis of this study. The authors introduce an innovative top-down selective chemical etching approach, employing HCl and HF as etching agents, yielding distinct outcomes. HCl etching facilitates boron species extraction while eliminating aluminum constituents, whereas HF etching exhibits the opposite effect. This chemically-guided process capitalizes on boron's electron-deficient nature, dictating the etching mechanism. The resulting borophene nanosheets, derived from precursor AlB_2 through selective etching, demonstrate a strong affinity for polydopamine (PDA) surface modification. This enhancement enhances cellular uptake and biocompatibility, yielding a responsive and biocompatible delivery nano platform. The investigation employs comprehensive morphological analyses, uncovering hexagonal nanosheets and distinct borophene phases via high-resolution TEM. Photothermal studies reveal inherent material potential, further augmented by PDA functionalization, elucidating degradation dynamics. Mechanistic insights, encompassing endocytosis, ROS induction, and in vitro cytotoxicity assessments, highlight the material's therapeutic potential for light-induced tumor treatment. Borophene's multifaceted utility extends beyond cellular boundaries, serving as a versatile contrast agent for tumor visualization through photoacoustic and fluorescence imaging. In-depth in vivo studies validate its tumor-targeting efficacy, bolstered by thorough toxicity evaluations, positioning it as a promising candidate for translational applications [86].

4.4 Biosensing and Diagnostics

Biosensing and diagnostic platforms are pivotal in healthcare, especially for disease detection, monitoring, and management. These platforms seamlessly merge biological and chemical sensing components with cutting-edge technologies, facilitating swift, precise, and individualized diagnoses for diverse conditions. By offering early disease identification, tailored treatments, ongoing surveillance, and enhanced patient results, these versatile tools significantly impact a broad spectrum of diseases and medical scenarios [87].

Boron nanosheets have been widely employed in biosensing/diagnostic platforms, especially for glucose. Glucose biosensors and monitoring devices are crucial tools that empower individuals with diabetes to take control of their health, make informed decisions, and work towards achieving optimal glucose control, ultimately leading to better overall well-being and a reduced risk of diabetes-related complications. They also play a crucial role in pre-diabetes management and can be integrated into telehealth for remote monitoring and guidance. The glucose-detecting capabilities of PAN: β-rhombohedral borophene-based non-enzymatic electrochemical biosensors were investigated in a recent work conducted by Taşaltın et al. The study begins by examining the complex structural characteristics of β-rhombohedral crystalline structured borophene, synthesized via physical exfoliation. The resulting combination of PAN and β borophene gives rise to the PAN: β-Rhombohedral Borophene nanocomposite, which is distinguished by its unique nanomorphology. The electrochemical biosensors used for glucose sensing exhibit two distinct redox peaks throughout a wide range of glucose concentrations, indicating the non-enzymatic character of the sensing mechanism. The PAN: β-Rhombohedral Borophene-based biosensor has exceptional sensitivity and an impressive charge transfer mechanism, resulting in enhanced glucose sensing capabilities even under ambient conditions. The results of the quantitative analyses reveal a sensitivity of 8.43 $\mu A\ mM^{-1}\ cm^{-2}$ during a voltammetry cycle lasting 1 min. Additionally, the limit of detection (LOD) is determined to be 0.099 mM. The interaction pathway between PAN and glucose is of notable importance, with a primary focus on the involvement of nitrogen sites on PAN and the mediation of electrical charge through glucose's –OH side group. Furthermore, the interaction between PAN: β-Rhombohedral borophene exploits the unfavorable zeta potential of borophene, resulting in enhanced redox activity and the emergence of unique oxidation–reduction peaks. The aforementioned research represents a significant breakthrough in the field, resulting in the development of a new biosensor known as PAN: β-Rhombohedral Borophene-based biosensor. This innovative technology demonstrates exceptional performance in terms of both sensitivity and stability while also obviating the requirement for enzyme immobilization [88] (Fig. 4a).

In a related work conducted by Baytemir et al., novel non-enzymatic electrochemical sensors were produced for glucose detection under ambient circumstances. These sensors were based on nickel phthalocyanine (NiPc) and a nanocomposite of NiPc and borophene and were specifically designed for this purpose. The electrical conductivities of the separate components, namely NiPc and the nanocomposite of NiPc-borophene, were measured and quantified systematically. The results indicate that the electrical conductivity of NiPc is 3×10^{-13} S cm^{-1}, while the nanocomposite of NiPc-borophene exhibits a higher electrical conductivity of 9.5×10^{-9} S cm^{-1}. The incorporation of borophene into the NiPc matrix presents itself as a crucial method for enhancing the electrical conductivity of NiPc. The enhancement in question is ascribed to the significant advantages in charge transfer provided by

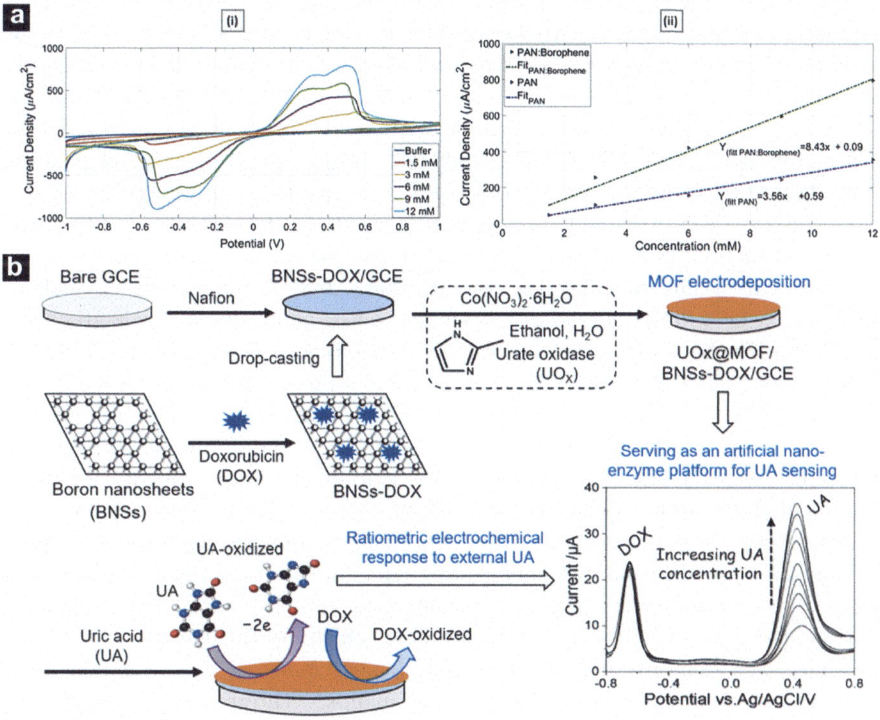

Fig. 4 **a** (i) Sensor response of PAN: β-rhombohedral borophene-based electrochemical biosensor against glucose. **b** (ii) Sensitivity graph of the PAN-based and PAN: β-Rhombohedral Borophene-based biosensors. Adapted with permission from [88], copyright Elsevier 2021. **b** Schematic illustration of the preparation of UOx@MOF/BNSs-DOX/GCE sensing platform and its application as a platform for enzymatic and ratiometric electrochemical biosensing of uric acid. Adapted with permission from [90], copyright Elsevier 2022

borophene, a characteristic that plays a crucial role in improving both the sensitivity of sensors and the limits of detection. The electrochemical sensors were subjected to a voltammetric cycle lasting 60 s to investigate glucose concentrations ranging from 1.5 to 24 mM. Remarkably, the sensor composed of a nanocomposite of NiPc-borophene showed superior performance compared to the sensor solely composed of NiPc. The former exhibited a significantly higher sensitivity of 10.31 μA mM^{-1} cm^{-2}, while the latter only achieved a sensitivity of 0.08 μA mM^{-1} cm^{-2}. The efficacy of this nanocomposite technique is further emphasized by the limits of detection, which are 3 μM and 0.15 μM for sensors based on NiPc and NiPc-borophene nanocomposites, respectively. The notable enhancement in sensitivity and detection characteristics is supported by the strategic application of borophene nanosheets, which are well-known for their remarkable chemical stability and exceptional carrier mobility [89].

Besides glucose, borophene-based biosensors have been used for uric acid, urea, and dopamine detection. In the study by Yang et al., BNSs are ingeniously employed to enhance the efficacy of a multifunctional nano-enzyme platform designed for enzymatic and ratiometric electrochemical biosensing of uric acid (UA). The BNSs are prepared through ultrasound-assisted liquid-phase exfoliation and subsequently combined with doxorubicin (DOX) to create a BNSs-DOX complex. This complex, drop-casted onto a glassy carbon electrode (GCE), forms the basis of a BNSs-DOX/GCE modified electrode. Notably, the study takes a step further by electrodepositing cobalt-based metal–organic framework (MOF) encapsulating urate oxidase (UOx) onto the BNSs-DOX-modified surface, thus fabricating a UOx@MOF/BNSs-DOX nanohybrid-modified GCE. This sophisticated nanohybrid-based electrode emerges as a potent nano-enzyme platform, orchestrating diverse functionalities such as DOX carrier, UOx immobilization, enzymatic redox, and ratiometric electrochemical sensing. The biosensor exhibits impressive analytical performance, detecting UA within a range of 0.1–200 μM with an exceptional limit of detection (LOD) of 0.025 μM (Fig. 4b). The judicious use of boron nanosheets within this intricate architecture not only bolsters sensitivity, selectivity, and stability but also signifies the potential of borophene-derived materials in elevating the capabilities of biosensing technologies for clinical applications, including early diagnosis of UA-related diseases [90].

Güngör et al. implemented the synthesis of nanocomposites consisting of copper phthalocyanine (CuPc) and borophene through the utilization of sonochemical techniques. By using the surface and chemical properties of the material, the researchers implemented it as a biosensor with remarkable selectivity for detecting urea. This selectivity was demonstrated even in complicated matrices that included other ions, such as glucose, fructose, lactose, and maltose. Through a clean and cost-effective sonication method, borophene nanosheets are synthesized and integrated into polyaniline (PANI). The resulting PANI/borophene nanocomposite material exhibits enhanced electrochemical performance, as evidenced by larger peak currents compared to the PANI-based sensor. This electrochemical synergy enabled the successful detection of dopamine within a concentration range of 0.15625–5 μM.

Remarkably, the PANI/borophene nanocomposite-based sensor showcases a significantly heightened sensitivity of 385.05 $\mu A \mu M^{-1}$ cm^{-2} and an impressive detection limit of 0.017 μM [91].

4.5 DNA Sequencing

DNA sequencing is a transformative molecular biology technique that enables the determination of the precise order of nucleotides (adenine, thymine, cytosine, and guanine) within a DNA molecule. By leveraging enzymatic reactions, fluorescent labels, and high-throughput technologies, DNA sequencing allows for the accurate reading of DNA sequences. This process plays a fundamental role in diverse scientific domains, including genomics, genetics, evolutionary biology, and biomedical research. It has led to unparalleled insights into genetic variation, disease susceptibility, molecular evolution, and the identification of genetic mutations underlying various health conditions [92]. DNA sequencing techniques encompass various techniques, including Sanger sequencing (first-gen) and Next-Generation Sequencing (NGS; second-gen), enabling the deciphering of DNA's nucleotide order. First-generation involves chain termination and gel electrophoresis, while NGS parallelizes sequencing for high throughput. Third-generation methods like PacBio and Oxford Nanopore offer long reads and real-time sequencing [93]. Borophene can theoretically be used in nanopore-based techniques, enhancing existing platforms, or developing sensitive DNA sensors. On this note, several studies have been reported.

The utilization of a borophene nanosheet for DNA sequencing is investigated in a recent work conducted by Rastgou et al. This exploration is carried out through the application of density functional theory calculations. This work examines the interplay between four nucleobases, namely adenine (A), guanine (G), thymine (T), and cytosine (C), and the borophene nanosheet. The investigation primarily focuses on the energetic and electronic sensitivities of these interactions. The researchers have discovered that the introduction of nucleobases onto the borophene nanosheet results in diverse alterations in its electrical conductivity, hence giving rise to discernible electrical signals. The sequence of reactivity of the borophene nanosheet towards nucleobases is as follows: adenine (A) exhibits the highest reactivity, followed by guanine (G), cytosine (C), and thymine (T). This discovery implies that borophene exhibits promising potential for utilization in devices specifically engineered for DNA sequencing. This study examines the interaction between individual nucleobases and borophene nanosheets, specifically investigating the formation of stable adsorption complexes and quantifying their respective interaction energies. It is worth mentioning that cytosine exerts the most pronounced influence on the conductivity of borophene, resulting in the most substantial electrical signal. Thymine follows suit, but with a lesser impact, whereas adenine and guanine produce comparatively weaker signals. The article additionally emphasizes the significance of the HOMO–LUMO gap in elucidating the relationship with electrical conductivity. Modifications occurring inside this void after the adsorption of nucleobases can induce variations in conductivity, hence giving rise to discernible electrical signals that may be employed to ascertain the sequence of DNA nucleobases. The findings indicate that the electrical sensitivity of the borophene nanosheet can be ranked as follows: C > T > A ~ G [94]. This study highlights the potential of borophene nanosheets in the field of DNA sequencing and offers valuable insights into their electronic and energetic sensitivities, which could be utilized in the advancement of sequencing devices.

5 Conclusion

2D boron nanosheets possess exceptional properties that render them highly appealing for diverse applications. Their ability to exhibit tunable bandgaps, anisotropic electrical conductivity, high thermal conductivity, and mechanical strength unlocks a wealth of opportunities in fields such as electronics, optoelectronics, and energy storage. Moreover, their catalytic activity, potential as drug carriers, and contrast agents in bioimaging showcase their versatility in biomedical applications. Nonetheless, a deeper exploration of the fundamental properties and behavior of these nanosheets is essential to fully harness the potential of 2D boron nanosheets and enable the development of tailored applications. It is imperative to conduct further research aimed at comprehending and optimizing synthesis methods while addressing challenges about scalability, stability, and toxicity. By sustaining continued investigation and development, these nanosheets can revolutionize various industries and contribute significantly to future technological advancements.

References

1. Bayda, S., Adeel, M., Tuccinardi, T., Cordani, M., Rizzolio, F.: The history of nanoscience and nanotechnology: from chemical–physical applications to nanomedicine. Molecules **25**, 112 (2020); **25**, 112 (2019). https://doi.org/10.3390/MOLECULES25010112
2. Rana, D., Salave, S., Longare, S., Agarwal, R., Kalia, K., Benival, D.: Nanotherapeutics in tumour microenvironment for cancer therapy. Nanosci. Nanotechnology-Asia **12** (2021). https://doi.org/10.2174/2210681211666210908144839
3. Singh, V., Yadav, P., Mishra, V.: Recent advances on classification, properties, synthesis, and characterization of nanomaterials. Green Synth. Nanomater. Bioenergy Appl. 83–97 (2020). https://doi.org/10.1002/9781119576785.CH3
4. Salave, S., Rana, D., Vitore, J., Jain, A.: Functionalized carbon nanotubes for cell tracking. Funct. Carbon Nanotub. Biomed. Appl. 319–338 (2023) https://doi.org/10.1002/9781119905080.CH13
5. Sangole, S., Salave, S., Rana, D., Shah, S., Medhe, T.P., Benival, D.: Electrospun nanofiber composite for levofloxacin in ocular drug delivery. Pharm. Nanotechnol. **10**, 393–400 (2022). https://doi.org/10.2174/2211738510666220902151906
6. Geim, A.K., Novoselov, K.S.: The rise of graphene. Nat. Mater. **63**, *6*, 183–191 (2007). https://doi.org/10.1038/nmat1849
7. Zhang, Q., Zhang, J., Wan, S., Wang, W., Fu, L.: Stimuli-responsive 2D materials beyond graphene. Adv. Funct. Mater. **28**, 1802500 (2018). https://doi.org/10.1002/ADFM.201802500

8. Lin, Y.C., Torsi, R., Younas, R., Hinkle, C.L., Rigosi, A.F., Hill, H.M., Zhang, K., Huang, S., Shuck, C.E., Chen, C., et al.: Recent advances in 2D material theory, synthesis, properties, and applications. ACS Nano (2022). https://doi.org/10.1021/ACSNANO.2C12759/ASSET/IMAGES/LARGE/NN2C12759_0025.JPEG
9. Mounet, N., Gibertini, M., Schwaller, P., Campi, D., Merkys, A., Marrazzo, A., Sohier, T., Castelli, I.E., Cepellotti, A., Pizzi, G., et al.: Two-dimensional materials from high-throughput computational exfoliation of experimentally known compounds. Nat. Nanotechnol. **133**, *13*, 246–252 (2018). https://doi.org/10.1038/s41565-017-0035-5
10. Barua, S., Sahu, D., Shahnaz, N., Khan, R.: Chemistry of two-dimensional nanomaterials. Two-Dimensional Nanostructures Biomed. Technol. A Bridg. between Mater. Sci. Bioeng. 1–33 (2020). https://doi.org/10.1016/B978-0-12-817650-4.00001-2
11. Zhang, R., Cheung, R., Zhang, R., Cheung, R.: Mechanical properties and applications of two-dimensional materials. Two-dimensional Mater. - Synth. Charact. Potential Appl. (2016). https://doi.org/10.5772/64017
12. Zhang, Z., Penev, E.S., Yakobson, B.I.: Two-Dimensional Boron: Structures, Properties and Applications. Chem. Soc. Rev. **46**, 6746–6763 (2017). https://doi.org/10.1039/C7CS00261K
13. Yang, M., Jin, H., Sun, Z., Gui, R.: Monoelemental two-dimensional boron nanomaterials beyond theoretical simulations: from experimental preparation, functionalized modification to practical applications. Adv. Colloid Interface Sci. **304** (2022). https://doi.org/10.1016/J.CIS.2022.102669
14. Duo, Y., Xie, Z., Wang, L., Mahmood Abbasi, N., Yang, T., Li, Z., Hu, G., Zhang, H.: Borophene-based biomedical applications: status and future challenges. Coord. Chem. Rev. **427** (2021). https://doi.org/10.1016/J.CCR.2020.213549
15. Lin, Y., Connell, J.W.: Advances in 2D boron nitride nanostructures: nanosheets, nanoribbons, nanomeshes, and hybrids with graphene. Nanoscale **4**, 6908–6939 (2012). https://doi.org/10.1039/C2NR32201C
16. Zhang, K., Feng, Y., Wang, F., Yang, Z., Wang, J.: Two dimensional hexagonal boron nitride (2D-HBN): synthesis, properties and applications. J. Mater. Chem. C **5**, 11992–12022 (2017). https://doi.org/10.1039/C7TC04300G
17. Roy, S., Zhang, X., Puthirath, A.B., Meiyazhagan, A., Bhattacharyya, S., Rahman, M.M., Babu, G., Susarla, S., Saju, S.K., Tran, M.K., et al.: Structure, properties and applications of two-dimensional hexagonal boron nitride. Adv. Mater. **33**, 2101589 (2021). https://doi.org/10.1002/ADMA.202101589
18. Ma, D., Wang, R., Zhao, J., Chen, Q., Wu, L., Li, D., Su, L., Jiang, X., Luo, Z., Ge, Y., et al.: A self-powered photodetector based on two-dimensional boron nanosheets. Nanoscale **12**, 5313–5323 (2020). https://doi.org/10.1039/D0NR00005A
19. John, S.K., Anappara, A.A.: Aqueous dispersions of highly luminescent boron-rich nanosheets by the exfoliation of polycrystalline titanium diboride. New J. Chem. **43**, 9953–9960 (2019). https://doi.org/10.1039/C9NJ01502G
20. Xu, J., Chang, Y., Gan, L., Ma, Y., Zhai, T., Xu, J.Q., Chang, Y.Y., Gan, L., Ma, Y., Zhai, T.Y.: Ultrathin single-crystalline boron nanosheets for enhanced electro-optical performances. Adv. Sci. **2**, 1500023 (2015). https://doi.org/10.1002/ADVS.201500023
21. Sun, Y., Gao, S., Lei, F., Xie, Y.: Atomically-thin two-dimensional sheets for understanding active sites in catalysis. Chem. Soc. Rev. **44**, 623–636 (2015). https://doi.org/10.1039/C4CS00236A
22. Xu, S.G., Li, X.T., Zhao, Y.J., Liao, J.H., Xu, W.P., Yang, X.B., Xu, H.: Two-dimensional semiconducting boron monolayers. J. Am. Chem. Soc. **139**, 17233–17236 (2017). https://doi.org/10.1021/JACS.7B08680/ASSET/IMAGES/LARGE/JA-2017-08680V_0003.JPEG
23. Mortazavi, B., Makaremi, M., Shahrokhi, M., Raeisi, M., Singh, C.V., Rabczuk, T., Pereira, L.F.C.: Borophene Hydride: a stiff 2D material with high thermal conductivity and attractive optical and electronic properties. Nanoscale **10**, 3759–3768 (2018). https://doi.org/10.1039/C7NR08725J
24. Yin, Y., Hu, Y., Feng, C., Li, S., Li, B.L., Li, D.F.: Strongly anisotropic thermal conductivity in planar hexagonal borophene oxide sheet. Phys. Lett. A **384**, 126457 (2020). https://doi.org/10.1016/J.PHYSLETA.2020.126457

25. Jiang, H.R., Lu, Z., Wu, M.C., Ciucci, F., Zhao, T.S.: Borophene: a promising anode material offering high specific capacity and high rate capability for lithium-ion batteries. Nano Energy **23**, 97–104 (2016). https://doi.org/10.1016/J.NANOEN.2016.03.013
26. Joshi, A., Tomar, A.K., Singh, G., Sharma, R.K.: Engineering oxygen defects in the boron nanosheet for stabilizing complex bonding structure: an approach for high-performance supercapacitor. Chem. Eng. J. **407**, 127122 (2021). https://doi.org/10.1016/J.CEJ.2020.127122
27. Zhan, C., Zhang, P., Dai, S., Jiang, D.E.: Boron supercapacitors. ACS Energy Lett. **1**, 1241–1246 (2016). https://doi.org/10.1021/ACSENERGYLETT.6B00483/ASSET/IMAGES/LARGE/NZ-2016-00483E_0006.JPEG
28. Angizi, S., Akbar, M.A., Darestani-Farahani, M., Kruse, P.: Review—two-dimensional boron carbon nitride: a comprehensive review. ECS J. Solid State Sci. Technol. **9**, 083004 (2020). https://doi.org/10.1149/2162-8777/ABB8EF
29. Aliofkhazraei, M., Ali, N.: AFM applications in micro/nanostructured coatings. Compr. Mater. Process. **7**, 191–241 (2014). https://doi.org/10.1016/B978-0-08-096532-1.00712-3
30. Ortiz, W., Ramirez, N.J., Barrionuevo, D., Bhattarai, M.K., Feng, P.: Characterization of 2D boron nitride nanosheets with hysteresis effect in the Schottky junctions. Nano Express **2**, 010020 (2021). https://doi.org/10.1088/2632-959X/ABDF6C
31. Main, K.H.S., Provan, J.I., Haynes, P.J., Wells, G., Hartley, J.A., Pyne, A.L.B.: Atomic force microscopy—a tool for structural and translational DNA research. APL Bioeng. **5**, 31504 (2021). https://doi.org/10.1063/5.0054294/149820
32. Ikram, M., Wakeel, M., Hassan, J., Haider, A., Naz, S., Ul-Hamid, A., Haider, J., Ali, S., Goumri-Said, S., Kanoun, M.B.: Impact of Bi doping into boron nitride nanosheets on electronic and optical properties using theoretical calculations and experiments. Nanoscale Res. Lett. **16**, 1–17 (2021). https://doi.org/10.1186/S11671-021-03542-X/FIGURES/11
33. Makuła, P., Pacia, M., Macyk, W.: How to correctly determine the band gap energy of modified semiconductor photocatalysts based on UV-vis spectra. J. Phys. Chem. Lett. **9**, 6814–6817 (2018). https://doi.org/10.1021/ACS.JPCLETT.8B02892/SUPPL_FILE/JZ8B02892_LIVESLIDES.MP4
34. Peak, D.: Fourier transform infrared spectroscopy. Encycl. Soils Environ. **4**, 80–85 (2005). https://doi.org/10.1016/B0-12-348530-4/00174-0
35. Guerra, V., Wan, C., Degirmenci, V., Sloan, J., Presvytis, D., McNally, T.: 2D boron nitride nanosheets (BNNS) prepared by high-pressure homogenisation: structure and morphology. Nanoscale **10**, 19469–19477 (2018). https://doi.org/10.1039/C8NR06429F
36. Shahabuddin, S., Khanam, R., Khalid, M., Sarih, N.M., Ching, J.J., Mohamad, S., Saidur, R.: Synthesis of 2D boron nitride doped polyaniline hybrid nanocomposites for photocatalytic degradation of carcinogenic dyes from aqueous solution. Arab. J. Chem. **11**, 1000–1016 (2018). https://doi.org/10.1016/J.ARABJC.2018.05.004
37. Titus, D., James Jebaseelan Samuel, E., Roopan, S.M.: Nanoparticle characterization techniques. Green Synth. Charact. Appl. Nanoparticles 303–319 (2019). https://doi.org/10.1016/B978-0-08-102579-6.00012-5
38. Yola, M.L., Atar, N.: A novel detection approach for serotonin by graphene quantum dots/two-dimensional (2D) hexagonal boron nitride nanosheets with molecularly imprinted polymer. Appl. Surf. Sci. **458**, 648–655 (2018). https://doi.org/10.1016/j.apsusc.2018.07.142
39. Shi, L., Ling, C., Ouyang, Y., Wang, J.: High intrinsic catalytic activity of two-dimensional boron monolayers for the hydrogen evolution reaction. Nanoscale **9**, 533–537 (2017). https://doi.org/10.1039/c6nr06621f
40. Kong, X., Huang, Y., Liu, Q.: Two-dimensional boron-doped graphyne nanosheet: a new metal-free catalyst for oxygen evolution reaction. Carbon N. Y. **123**, 558–564 (2017). https://doi.org/10.1016/j.carbon.2017.08.003
41. Li, J., Gan, L., Liu, Y., Mateti, S., Lei, W., Chen, Y., Yang, J.: Boron nitride nanosheets reinforced waterborne polyurethane coatings for improving corrosion resistance and antifriction properties. Eur. Polym. J. **104**, 57–63 (2018). https://doi.org/10.1016/J.EURPOLYMJ.2018.04.042

42. Wu, M., Zhou, Y., Zhang, H., Liao, W.: 2D boron nitride nanosheets for smart thermal management and advanced dielectrics. Adv. Mater. Interfaces **9**, 2200610 (2022). https://doi.org/10.1002/ADMI.202200610
43. Salave, S., Rana, D., Benival, D.: Peptide functionalised nanocarriers for bone specific delivery of PTH (1–34) in osteoporosis. Curr. Nanomedicine **11**, 142–148 (2021). https://doi.org/10.2174/2468187312666211220112324
44. Rana, D., Salave, S., Jain, S., Shah, R., Benival, D.: Systematic development and optimization of teriparatide-loaded nanoliposomes employing quality by design approach for osteoporosis. J. Pharm. Innov. 1–15 (2022). https://doi.org/10.1007/S12247-022-09663-9/METRICS
45. Rana, D., Salave, S., Benival, D.: Emerging trends in abuse-deterrent formulations: technological insights and regulatory considerations. Curr. Drug Deliv. **19**, 846–859 (2021). https://doi.org/10.2174/1567201818666211208101035
46. Gupta, R., Salave, S., Rana, D., Karunakaran, B., Butreddy, A., Benival, D., Kommineni, N., Mousavifar, L., Gupta, R., Salave, S., et al.: Versatility of liposomes for antisense oligonucleotide delivery: a special focus on various therapeutic areas. Pharm. **15**, 1435 (2023); **15**, 1435 (2023). https://doi.org/10.3390/PHARMACEUTICS15051435
47. Khunt, D., Prajapati, B.G., Prajapti, M., Misra, M., Salave, S., Patel, J.K., Patel, R.J.: Drug delivery by micro, nanoemulsions in tuberculosis. Tubercular Drug Deliv. Syst. 173–188 (2023). https://doi.org/10.1007/978-3-031-14100-3_9
48. Salave, S., Shinde, S.D., Rana, D., Sahu, B., Kumar, H., Patel, R., Benival, D., Kommineni, N.: Peptide engraftment on pegylated nanoliposomes for bone specific delivery of PTH (1–34) in osteoporosis. Pharmaceutics **15**, 608 (2023). https://doi.org/10.3390/PHARMACEUTICS15020608/S1
49. Salave, S., Prayag, K., Rana, D., Amate, P., Pardhe, R., Jadhav, A., Jindal, A.B., Benival, D.: Recent progress in hot melt extrusion technology in pharmaceutical dosage form design. Recent Adv. Drug Deliv. Formul. **16**, 170–191 (2022). https://doi.org/10.2174/2667387816666220819124605
50. Anton, N., Desai, N., Rana, D., Pande, S., Salave, S., Giri, J., Benival, D., Kommineni, N.: "bioinspired" membrane-coated nanosystems in cancer theranostics: a comprehensive review. Pharm. **15**, 1677 (2023); **15**, 1677 (2023). https://doi.org/10.3390/PHARMACEUTICS15061677
51. Salave, S., Rana, D., Bodar, A., Khunt, D., Prajapati, B., Patel, J.: Viral infections: current treatment options. Viral Drug Deliv. Syst. 65–89 (2023). https://doi.org/10.1007/978-3-031-20537-8_4
52. Vitore, J.G., Bharathi, K., Salave, S., Rana, D., Perla, A., Gupta, S., Shah, S., Pardhe, R., Chittemreddy, P., Kashid, S., et al.: Intranasal transmucosal drug delivery: an alternative approach to the parenteral route for medical emergencies. J. Drug Deliv. Sci. Technol. **83**, 104421 (2023). https://doi.org/10.1016/J.JDDST.2023.104421
53. Gupta, S., Perla, A., Roy, A., Vitore, J.G., Bharathi, K., Salave, S., Rana, D., Sharma, A., Rathod, R., Kumar, H., et al.: In vivo evaluation of almotriptan malate formulation through intranasal route for the treatment of migraine: systematic development and pharmacokinetic assessment. AAPS PharmSciTech **24**, 1–13 (2023). https://doi.org/10.1208/S12249-022-02496-2/METRICS
54. Rawat, G., Kolhe, S., Rana, D., Salave, S., Benival, D.: Exploring the therapeutic potential of cyclosporine for ophthalmic indications by novel carrier systems. Crit. Rev. Ther. Drug Carr. Syst. **40**, 1–45 (2023). https://doi.org/10.1615/CRITREVTHERDRUGCARRIERSYST.2022043085
55. Salave, S., Rana, D., Pardhe, R., Bule, P., Benival, D.: Unravelling micro and nano vesicular system in intranasal drug delivery for epilepsy. Pharm. Nanotechnol. **10**, 182–193 (2022). https://doi.org/10.2174/2211738510666220426115340
56. Agarwal, R., Rana, D., Salave, S., Benival, D.: Dexamethasone loaded electrospun nanocomposite ocular insert: in-vitro drug release and mechanical assessment. Curr. Nanomedicine **12**, 150–158 (2022). https://doi.org/10.2174/2468187312666220806133901

57. Rana, D., Gupta, R., Bharathi, K., Pardhe, R., Jain, N.K., Salave, S., Prasad, R., Benival, D., Kommineni, N.: Porous silica nanoparticles for targeted bio-imaging and drug delivery applications. Nanomater. Healthc. 133–154 (2023)
58. Rana, D., Desai, N., Salave, S., Karunakaran, B., Giri, J., Benival, D., Gorantla, S., Kommineni, N.: Collagen-based hydrogels for the eye: a comprehensive review. Gels **9**, 643 (2023). https://doi.org/10.3390/GELS9080643
59. Salave, S., Rana, D., Prayag, K., Shah, S., Rawat, G., Sharma, N., Jindal, A.B., Patel, R., Benival, D.: Recent advances in teriparatide delivery by-virtue-of novel drug delivery approaches for the management of osteoporosis. Crit. Rev. Ther. Drug Carr. Syst. **40**, 93–113 (2023). https://doi.org/10.1615/CRITREVTHERDRUGCARRIERSYST.2023045014
60. Salave, S., Rana, D., Sharma, A., Bharathi, K., Gupta, R., Khode, S., Benival, D., Kommineni, N.: Polysaccharide based implantable drug delivery: development strategies, regulatory requirements, and future perspectives. Polysaccharides 2022 **2022**, *3*, 625–654, https://doi.org/10.3390/POLYSACCHARIDES3030037
61. Desai, N., Rana, D., Salave, S., Gupta, R., Patel, P., Karunakaran, B., Sharma, A., Giri, J., Benival, D., Kommineni, N. Chitosan: A potential biopolymer in drug delivery and biomedical applications. Pharm. **15**, 1313 (2023). https://doi.org/10.3390/PHARMACEUTICS15041313
62. Karunakaran, B., Gupta, R., Patel, P., Salave, S., Sharma, A., Desai, D., Benival, D., Kommineni, N.: Emerging trends in lipid-based vaccine delivery: a special focus on developmental strategies, fabrication methods, and applications. Vaccines **11**, 661 (2023). https://doi.org/10.3390/VACCINES11030661
63. Rana, D., Salave, S., Rawat, G., Benival, D.: Recent trends in drug delivery and emerging biomedical applications of gelatin for ophthalmic indications. Macromol. Res. **30**, 687–702 (2022). https://doi.org/10.1007/S13233-022-0078-9
64. Salave, S., Rana, D., Benival, D.: Encapsulation of anabolic peptide in lipid nano vesicles for osteoporosis. Curr. Protein Pept. Sci. **23**, 495–503 (2022). https://doi.org/10.2174/1389203723666220822120305
65. Salave, S., Jain, S., Shah, R., Benival, D.: Quantification of anti-osteoporotic anabolic peptide in stealth lipid nanovesicles through validated RP-HPLC method. J. AOAC Int. **106**, 40–48 (2022). https://doi.org/10.1093/JAOACINT/QSAC096
66. Salave, S., Rana, D., Kumar, H., Kommineni, N., Benival, D.: Anabolic peptide-enriched stealth nanoliposomes for effective anti-osteoporotic therapy. Pharmaceutics **14**, 2417 (2022). https://doi.org/10.3390/PHARMACEUTICS14112417/S1
67. Salave, S., Rana, D., Benival, D.: Dual targeting anti-osteoporotic therapy through potential nanotherapeutic approaches. Pharm. Nanotechnol. **10**, 384–392 (2022). https://doi.org/10.2174/2211738510666220902124653
68. Jindal, A.B., Bhide, A.R., Salave, S., Rana, D., Benival, D.: Long-acting parenteral drug delivery systems for the treatment of chronic diseases. Adv. Drug Deliv. Rev. **198**, 114862 (2023). https://doi.org/10.1016/J.ADDR.2023.114862
69. Vatanparast, M., Shariatinia, Z.: Hexagonal boron nitride nanosheet as novel drug delivery system for anticancer drugs: insights from DFT calculations and molecular dynamics simulations. J. Mol. Graph. Model. **89**, 50–59 (2019). https://doi.org/10.1016/J.JMGM.2019.02.012
70. Duverger, E., Balme, S., Bechelany, M., Miele, P., Picaud, F.: Natural payload delivery of the doxorubicin anticancer drug from boron nitride oxide nanosheets. Appl. Surf. Sci. **475**, 666–675 (2019). https://doi.org/10.1016/J.APSUSC.2018.12.273
71. Duverger, E., Picaud, F.: Theoretical study of ciprofloxacin antibiotic trapping on graphene or boron nitride oxide nanoflakes. J. Mol. Model. **26**, 1–10 (2020). https://doi.org/10.1007/S00894-020-04410-8/METRICS
72. Li, L., Dai, K., Li, J., Shi, Y., Zhang, Z., Liu, T., Jun, X.; Ruiping, Z., Liu, Z.: A boron-10 nitride nanosheet for combinational boron neutron capture therapy and chemotherapy of tumor. Biomaterials **268**, 120587 (2021). https://doi.org/10.1016/J.BIOMATERIALS.2020.120587
73. Cheng, Y., Han, Y., Zhang, W., Zeng, L., Long, Y., Wang, S., Weng, Q.: Gram-scale synthesis of boron nitride nanosheets by salt-template method for anticancer drug delivery. Chem. Eng. J. **437**, 135304 (2022). https://doi.org/10.1016/J.CEJ.2022.135304

74. Cheng, X., Li, F., Liang, L.: Boron neutron capture therapy: clinical application and research progress. Curr. Oncol. **29**, 7868–7886 (2022). https://doi.org/10.3390/CURRONCOL29100622
75. Loap, P., De Marzi, L., Almeida, C.E., Barcellini, A., Bradley, J., de Santis, M.C., Dendale, R., Jimenez, R., Orlandi, E., Kirova, Y.: Hadrontherapy techniques for breast cancer. Crit. Rev. Oncol. Hematol. **169**, 103574 (2022). https://doi.org/10.1016/J.CRITREVONC.2021.103574
76. Qi, P., Chen, Q., Tu, D., Yao, S., Zhang, Y., Wang, J., Xie, C., Pan, C., Peng, H.: The potential role of borophene as a radiosensitizer in boron neutron capture therapy (BNCT) and particle therapy (PT). Biomater. Sci. **8**, 2778–2785 (2020). https://doi.org/10.1039/D0BM00318B
77. Zhi, D., Yang, T., O'Hagan, J., Zhang, S., Donnelly, R.F.: Photothermal therapy. J. Control. Release **325**, 52–71 (2020). https://doi.org/10.1016/J.JCONREL.2020.06.032
78. Pham, T.C., Nguyen, V.N., Choi, Y., Lee, S., Yoon, J.: Recent strategies to develop innovative photosensitizers for enhanced photodynamic therapy. Chem. Rev. **121**, 13454–13619 (2021). https://doi.org/10.1021/ACS.CHEMREV.1C00381/ASSET/IMAGES/MEDIUM/CR1C00381_0262.GIF
79. Kang, Y., Ji, X., Li, Z., Su, Z., Zhang, S.: Boron-Based nanosheets for combined cancer photothermal and photodynamic therapy. J. Mater. Chem. B **8**, 4609–4619 (2020). https://doi.org/10.1039/D0TB00070A
80. Fu, Z., Williams, G.R., Niu, S., Wu, J., Gao, F., Zhang, X., Yang, Y., Li, Y., Zhu, L.M.: Functionalized boron nanosheets as an intelligent nanoplatform for synergistic low-temperature photothermal therapy and chemotherapy. Nanoscale **12**, 14739–14750 (2020). https://doi.org/10.1039/D0NR02291H
81. Sun, Z., Fan, T., Liu, Q., Huang, L., Hu, W., Shi, L., Wu, Z., Yang, Q., Liu, L., Zhang, H.: Autologous tumor antigens and boron nanosheet-based nanovaccines for enhanced photo-immunotherapy against immune desert tumors. Nanophotonics **10**, 2519–2535 (2021). https://doi.org/10.1515/NANOPH-2021-0229/ASSET/GRAPHIC/J_NANOPH-2021-0229_FIG_006.JPG
82. Xiao, X., Zhao, Y., Ma, P., Cheng, Z., Lin, J.: Boron-based nanosheets for ultrasound-mediated synergistic cancer therapy. Chem. Eng. J. **440**, 135812 (2022). https://doi.org/10.1016/J.CEJ.2022.135812
83. Dabhade, P., Bagde, A., Syed, Z.Q., Fulzele, P.: Utilization of nanomaterials in bioimaging and cancer theranostics. IOP Conf. Ser. Mater. Sci. Eng. **1259**, 012047 (2022). https://doi.org/10.1088/1757-899X/1259/1/012047
84. Ma, D., Zhao, J., Xie, J., Zhang, F., Wang, R., Wu, L., Liang, W., Li, D., Ge, Y., Li, J., et al.: Ultrathin boron nanosheets as an emerging two-dimensional photoluminescence material for bioimaging. Nanoscale Horiz. **5**, 705–713 (2020). https://doi.org/10.1039/C9NH00698B
85. Ji, X., Kong, N., Wang, J., Li, W., Xiao, Y., Gan, S.T., Zhang, Y., Li, Y., Song, X., Xiong, Q., et al.: A novel top-down synthesis of ultrathin 2D boron nanosheets for multimodal imaging-guided cancer therapy. Adv. Mater. **30**, 1803031 (2018). https://doi.org/10.1002/ADMA.201803031
86. Xie, Z., Duo, Y., Fan, T., Zhu, Y., Feng, S., Li, C., Guo, H., Ge, Y., Ahmed, S., Huang, W., et al.: Light-induced tumor theranostics based on chemical-exfoliated borophene. Light Sci. Appl. **11**, 1–14 (2022). https://doi.org/10.1038/s41377-022-00980-9
87. Patel, S., Nanda, R., Sahoo, S., Mohapatra, E.: Biosensors in health care: the milestones achieved in their development towards lab-on-chip-analysis. Biochem. Res. Int. (2016). https://doi.org/10.1155/2016/3130469
88. Taşaltın, C.: Glucose sensing performance of PAN: β-rhombohedral borophene based non-enzymatic electrochemical biosensor. Inorg. Chem. Commun. **133**, 108973 (2021). https://doi.org/10.1016/J.INOCHE.2021.108973
89. Baytemir, G., Gürol, İ, Karakuş, S., Taşaltın, C., Taşaltın, N.: Nickel phthalocyanine-borophene nanocomposite-based electrodes for non-enzymatic electrochemical detection of glucose. J. Mater. Sci. Mater. Electron. **33**, 16586–16596 (2022). https://doi.org/10.1007/S10854-022-08551-9/METRICS

90. Yang, M., Sun, Z., Jin, H., Gui, R.: Urate oxidase-loaded MOF electrodeposited on boron nanosheet-doxorubicin complex as multifunctional nano-enzyme platform for enzymatic and ratiometric electrochemical biosensing. Talanta **243**, 123359 (2022). https://doi.org/10.1016/J.TALANTA.2022.123359
91. Baytemir, G.: A non-enzymatic electrochemical sensor based on polyaniline/borophene nanocomposites for dopamine detection. Appl. Phys. A Mater. Sci. Process. **129**, 1–9 (2023). https://doi.org/10.1007/S00339-022-06364-5/METRICS
92. Shendure, J., Aiden, E.L.: The expanding scope of DNA sequencing. Nat. Biotechnol. 3011; **30**, 1084–1094 (2012). https://doi.org/10.1038/nbt.2421
93. Kumar, K.R., Cowley, M.J., Davis, R.L.: Next-generation sequencing and emerging technologies. Semin. Thromb. Hemost. **45**, 661–673 (2019). https://doi.org/10.1055/S-0039-1688446/ID/JR02634-12/BIB
94. Rastgou, A., Soleymanabadi, H., Bodaghi, A.: DNA sequencing by borophene nanosheet via an electronic response: a theoretical study. Microelectron. Eng. **169**, 9–15 (2017). https://doi.org/10.1016/J.MEE.2016.11.012

Potential Advancements and Upcoming Smart Research in Borophene: Challenges to Future

Mohd. Abubakar Sadique, Shalu Yadav, Pushpesh Ranjan, and Raju Khan

Abstract Borophene, a 2D form of boron, has attracted much interest because of its outstanding physical, chemical, and electrical characteristics, making it a very promising material for future technological progress. This chapter explores the possible progress and development of borophene, with a specific focus on its synthesis, characteristics, and uses. We investigate novel manufacturing methods that address existing obstacles in synthesis, such as the production of extensive and top-notch borophene sheets. The discussion encompasses the incorporation of borophene into advanced electronics, energy storage devices, and sensors, emphasizing its exceptional conductivity, flexibility, and large surface area. In addition, the chapter discusses the main obstacles that may hinder future advancements, such as concerns over stability, the ability to scale up production, and the requirement for thorough theoretical models to forecast the behavior of borophene in different settings. This chapter intends to give a roadmap for future exploration and development in borophene technology by presenting a comprehensive overview of current research trends and potential solutions to the issues. The purpose of this chapter is to provide valuable knowledge that will assist researchers in effectively utilizing the whole capabilities of borophene, therefore facilitating significant advancements in multiple influential domains.

Keywords Borophene · 2D nanomaterials · Synthesis techniques · Energy storage · Sensors · Stability · Biomedical · Future advancements

Mohd. A. Sadique · S. Yadav · P. Ranjan · R. Khan (✉)
CSIR—Advanced Materials and Processes Research Institute (AMPRI), Bhopal 462026, India
e-mail: khan.raju@ampri.res.in

Academy of Scientific and Innovative Research (AcSIR), Ghaziabad 201002, India

R. Khan et al. (eds.), *2D Boron Nanosheets*, Engineering Materials,
https://doi.org/10.1007/978-981-97-7014-4_9

1 Emergence of Borophene

Recently, bilayer (BL) and few-layer graphene and transition metal dichalcogenides in well-controlled stacking geometries have attracted immense interest due to their exotic and finely tunable physical properties. Unlike these natural layered materials that can be exfoliated and integrated mechanically due to interlayer van der Waals (vdW) interactions, synthetic two-dimensional (2D) materials have no bulk counterparts and thus form layered structures potentially via non-vdW interactions. Very recently, the first BL synthetic 2D boron (i.e., borophene) was realized on Ag(111) substrates with enhanced stability and promising computationally predicted properties such as superconductivity, antiferromagnetism, double Dirac cones, and nodal line fermions [1]. Borophene is the boron atom analog of graphene with interesting properties suitable for a wide variety of applications. Borophene (2D boron sheet) is a new type of 2D material, which is electron-deficient and cannot form graphene-like structures [2]. Following this approach, several synthetic 2D materials have been realized in recent years (for example, silicene, germanene, stanene, antimonene, bismuthine, and telluride). Though the study of these materials is still at an early stage, predictions and preliminary measurements of their properties confirm that they are complementary to conventional (that is, layered bulk-derived) 2D materials, as summarized in several recent reviews [3].

Borophene is a crystalline atomic monolayer of boron atoms, arranged in a hexagonal lattice structure similar to graphene. This arrangement results in a two-dimensional material with unique properties. It can be considered a new allotrope of boron, distinct from the more commonly encountered elemental boron [4, 5]. Since its theoretical prediction in 2014, borophene has engrossed researchers around the globe. Due to its distinctive configuration, which bears a resemblance to a honeycomb lattice, it possesses exceptional electronic and physical characteristics that offer tremendous potential for diverse technological progressions. Borophene exhibits remarkable in-plane elasticity and ideal strength [6]. It is theoretically predicted to be stronger than graphene, and more flexible in some configurations [7]. These properties make it a promising material for applications where low weight and high resilience are crucial, such as in nanoelectronics and aerospace engineering [8]. Under certain conditions, borophene has been shown to exhibit superconductivity [9, 10]. This means it can conduct electricity with zero resistance at very low temperatures [11]. This discovery has opened up exciting possibilities for its use in high-efficiency electronics and quantum technologies [12]. Depending on the specific arrangement of boron atoms in the lattice, borophene can exhibit either semiconducting or metallic properties [13, 14]. This tunability allows for tailoring its electronic behavior for specific applications, such as transistors, sensors, and optoelectronic devices [15, 16]. Borophene possesses a high density of Dirac electrons, which are massless particles that behave like relativistic particles [17, 18]. This property gives rise to unique electronic and transport properties, making it a promising material for next-generation electronics and spintronics [19, 20]. The properties of borophene are highly dependent on the direction of measurement

[21, 22]. This means that its conductivity, strength, and other characteristics can vary significantly depending on the angle at which it is probed [23, 24]. This anisotropy needs to be considered when designing devices and applications based on borophene [25]. Unique atomic arrangement and electronic characteristics in Borophene provide exceptional stability, customizable electronic properties, high carrier mobility, and a large surface area for interaction with other materials or molecules, making it promising for high-speed electronic devices, sensing, and catalytic applications due to its potential for high-speed devices.

Borophene is relatively stable under ambient conditions [26]. However, it can be susceptible to oxidation at higher temperatures [27]. Borophene exhibits low thermal conductivity, which can be advantageous for applications where heat dissipation is important [28, 29]. The large surface area of borophene makes it a promising candidate for applications like catalysis, energy storage, and sensors [30–32]. Borophene belongs to a new class of 2D materials beyond graphene, offering a distinct combination of strength, flexibility, and electronic properties [33]. This opens doors for exploring uncharted territory in materials science and engineering. Unlike many materials, the electronic properties of Borophene can be tailored from semiconducting to metallic by adjusting its atomic structure. This versatility allows for designing materials for specific applications, unlike the "one-size-fits-all" approach with other materials. The presence of high-density Dirac fermions in borophene grants it exotic electronic behavior, enabling applications in quantum computing, spintronics, and high-speed transistors [34]. The unique atomic arrangement and high surface area of Borophene make it a promising candidate for catalyzing various chemical reactions, potentially leading to more efficient and sustainable processes [35]. Borophene exhibits exceptional sensitivity to different stimuli, including light, gases, and molecules [31]. This paves the way for developing highly accurate and miniaturized sensors for various fields, from environmental monitoring to healthcare. Borophene-based transistors could be faster, more energy-efficient, and even flexible, revolutionizing electronics miniaturization and performance [36]. Harnessing the spin of Dirac fermions in borophene could lead to low-power spintronic devices with superior functionalities compared to conventional electronics. Borophene electrodes could enable high-performance batteries and supercapacitors with improved energy density and faster charging times [37]. Borophene-based catalysts could accelerate various chemical reactions, leading to more efficient production of fuels, chemicals, and pharmaceuticals. The sensitivity of Borophene to different stimuli opens doors for developing gas sensors for environmental monitoring, biosensors for disease detection, and even sensors for detecting explosives and harmful chemicals [38]. Unique properties of Borophene could be harnessed for building components of quantum computers, potentially leading to breakthroughs in computing power and materials science. This chapter delves into the exciting world of borophene, exploring its properties, applications, current research challenges, and future directions, paving the way for "smart" research endeavors (Fig. 1).

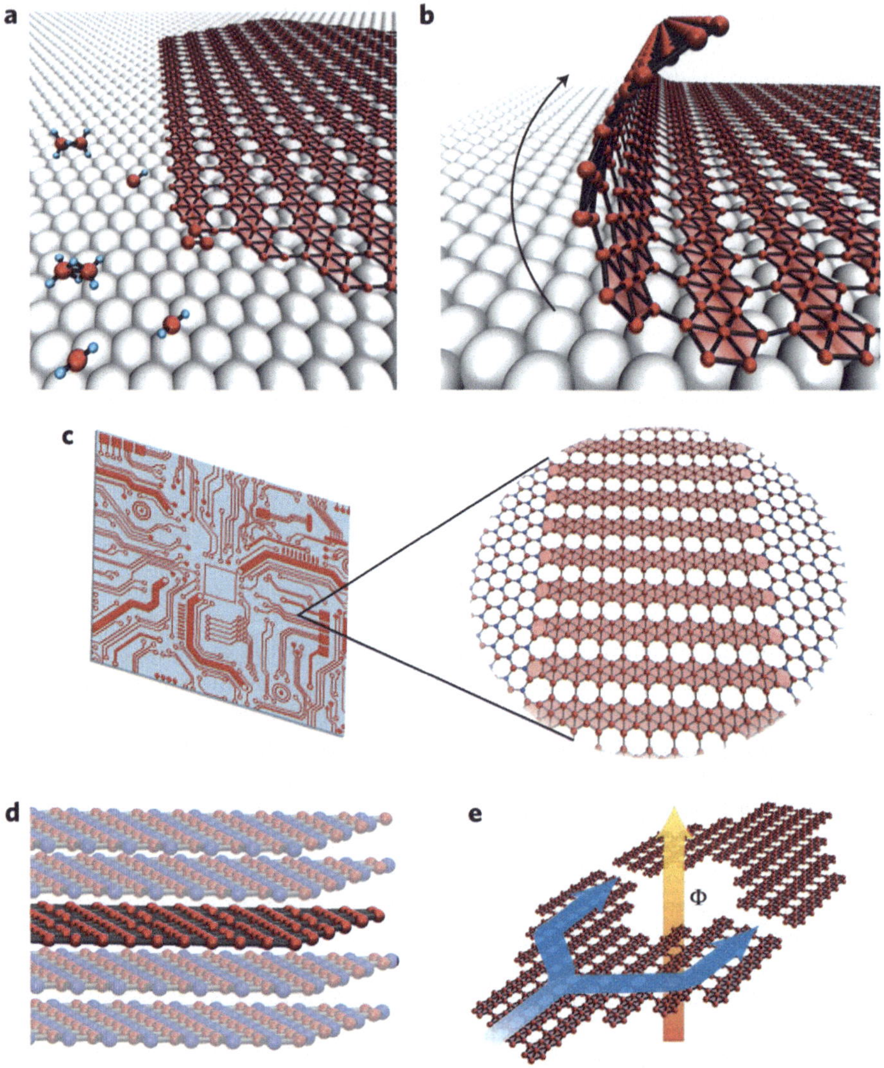

Fig. 1 **a** Chemical vapor deposition (CVD) uses diborane as a precursor for extensive domain formation. **b** Borophene is transferred from the substrate. **c** Forming lateral borophene-hBN heterostructures for electronic circuits. **d** Van der Waals heterostructure is used to encapsulate borophene. **e** Illustration shows a superconducting quantum interference device (SQUID) made of borophene superconducting leads. The symbol Φ represents the quantized magnetic flux that is enclosed within the SQUID. Reprinted (adapted) with permission from [3]. Copyright 2018 Springer Nature

2 Previous Research and Advancements in Borophene

Research on borophene has led to successful synthesis, detailed characterization, and initial demonstrations of its potential in various applications. Techniques like physical vapor deposition, chemical vapor deposition, and molecular beam epitaxy have been used [39–44]. Early theoretical predictions; the 1990s-2010s predicted borophene structures with unique electronic and physical properties, with honeycomb structure being the most stable [45, 46]. High reactivity and non-planar structure formation debated synthesis feasibility. A Berkeley team synthesized the first atomically thin borophene layer on a silver substrate using chemical vapor deposition, sparking research on molecular beam epitaxy, liquid metal exfoliation, and surface confinement [47]. Research on borophene structures, synthesis phases, substrate influence, and atomic defects has revealed properties like superconductivity, high electron mobility, and tunable bandgaps, depending on the structure and atomic arrangement [48–50].

From 2020 to the present, the focus has been on the characterization and applications of Borophene beyond its synthesis. Advanced characterization techniques like scanning tunneling microscopy and atomic resolution transmission electron microscopy have been employed to unravel the atomic-level details of borophene structures and their impact on properties [51, 52]. Researchers are actively exploring potential applications of borophene, including, high-performance transistors and flexible electronics, spintronic devices with low power consumption, and enhanced functionality [53]. Borophene is used in the development of next-generation batteries and supercapacitors with higher energy density and faster charging [54, 55]. Highly sensitive sensors are made out of Borophene for gas detection, environmental monitoring, biosensing, and even catalytic materials for efficient and sustainable chemical reactions [56, 57]. The focus of research in Borophene is shifting toward scalable and cost-effective synthesis techniques for large-scale production and practical applications. Researchers are exploring strategies to improve the stability of borophene against oxidation and environmental degradation [32]. Integrating borophene with other materials to create hybrid structures with synergistic properties is a promising avenue [36, 58]. Theoretical simulations are guiding the design of novel borophene structures with tailor-made properties for specific applications.

3 Current Challenges in Borophene Research

Borophene research has several obstacles, including limited synthesis, transferability, stability under ambient settings, and understanding of structure–property connections. Current technologies produce less quantity, necessitating additional research for improved stability and targeted optimization. Additional study is required to improve the practical applicability of borophene.

3.1 Synthesis and Manufacturing Challenges

Borophene is typically synthesized on metal substrates using complex and expensive techniques like chemical vapor deposition (CVD) and molecular beam epitaxy (MBE) [26, 59–61]. These methods often result in small-scale, polycrystalline borophene films with limited control over the structure and properties [62]. Developing scalable and cost-effective methods for producing large-area, single-crystal borophene films is crucial for its practical applications.

Nanostructured boron allotropes have received relatively little experimental attention to date, in part because the precursors (such as diborane) that are usually employed are expensive and hazardous [63]. However, numerous theoretical studies have examined two-dimensional (2D) boron sheets i.e., borophene [2]. Tip-enhanced Raman spectroscopy (TERS) was used to study the chemical properties of bilayer-produced two-dimensional boron (borophene) at the atomic level. TERS analyzed interlayer B–B bond vibrations to determine the chemical properties of bilayer borophene. The study showed the chemical stability of bilayer borophene under controlled oxidizing conditions in a vacuum, highlighting the sensitivity of single-bond (B–O) reactions [1] (Fig. 2). The growth of boron monolayer is very difficult in experiments due to the high melting point of the boron element. Based on extensive theoretical and computational studies, two groups independently reported the successful synthesis of borophene on the Ag (110) surface by molecular beam epitaxy (MBE) [43, 64]. This breakthrough indicates a new family member of 2D materials is born, and the nanodevices based on boron elements could be possible. Then, borophene sheets with different phases and borophene nanoribbons have been experimentally synthesized on the substrates of Ag(111), Cu(111), Al(111), Ir(111), and Au(111) [40, 65–68]. However, experimental research on borophene is still lacking, and the investigations including growth, characterization, and applications are still awaiting, which should be the prerequisite for the application of borophene in the next-generation microelectronics [48].

3.2 Structural Stability and Electronic Properties Challenges

For newly proposed 2D materials, stability is an important aspect of experimental realization and large-scale production. As a relevantly new 2D material, borophene has been grown successfully recently on single crystal Ag substrates. Three main structures have been proposed (χ_3,$\beta_{12,}$ and striped borophene). However, the stability of the three structures is still under debate. Using first-principles calculations, the dynamical, thermodynamical, and mechanical stability of striped borophene is examined. Free-standing borophene is dynamically, thermodynamically, and mechanically stable, while striped borophene is dynamically and thermodynamically unstable due to high stiffness along a direction [69, 70]. The origin of high stiffness and high instability in striped borophene along a direction can both be attributed to strong

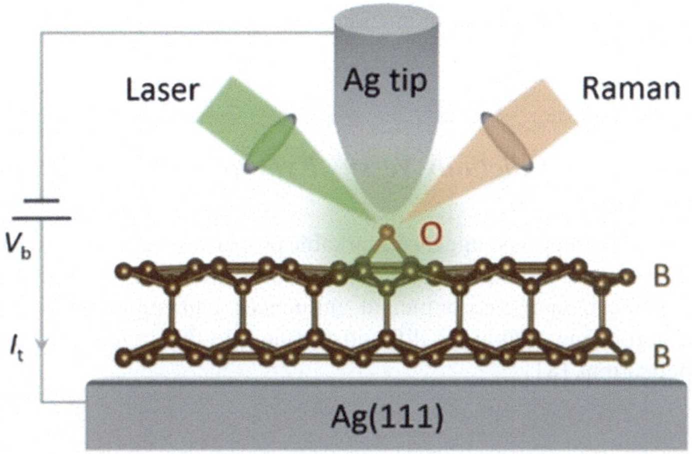

Fig. 2 Utilizing tip-enhanced Raman spectroscopy (TERS) provided atomic-scale spatial resolution in the investigation of the local chemical characteristics of bilayer synthesized two-dimensional boron (i.e., borophene). Reprinted (adapted) with permission from [1]. Copyright 2023 John Wiley & Sons

directional bonding [22, 71]. The studies show a deep connection between stability and strength and help researchers to estimate accurately the mechanical performance of 2D materials. However, the drawback of borophene is the shortage of surface passivation, which leads to unstable structure as demonstrated by previous studies [69, 72]. Due to this reason, a high vacuum atmosphere is required for carrying out experiments on borophene, since the chemically active surface of borophene may get affected (oxidized) by the environment [73]. Therefore, surface passivation with hydrogen is strictly required to change the metallic properties of borophene and also to stabilize the nanostructure. Although a series of 2D boron sheets or borophenes could be predicted by theorists, the borophenes in free-standing conditions showed instability in the dynamical calculations [27]. As a result, the borophenes observed in the STM are often partially oxidized at the edge of borophenes or protected by a Si/SiO_x buffer layer. Another theoretical attempt to enhance structural stability was performed by Xu et al. [74] through the surface hydrogenation of the borophene. The hydrogen attached to the surface does not change the structures of δ6-borophene but effectively improves the stability of the borophene.

3.3 Characterization and Manipulation Challenges

Borophene exists in various structural polymorphs, each with unique electronic and physical properties. Accurately characterizing and distinguishing these polymorphs poses a significant challenge [75–77]. Advanced characterization techniques, such as atomic resolution transmission electron microscopy (TEM) and scanning tunneling microscopy (STM), are essential for elucidating the structure–property relationships in borophene [78–80].

3.4 Theoretical Predictions Versus Experimental Realization

While theoretical predictions suggest remarkable properties for borophene, achieving these properties in actual experiments remains a challenge. Factors like substrate interaction, defects, and environmental influences can significantly affect the behavior of borophene, making it difficult to translate theoretical predictions into practical applications [81].

4 Potential Advancements in Borophene Research

To overcome the challenges mentioned in the previous section and unlock the full potential of borophene, research efforts are directed towards.

4.1 Novel Synthesis Techniques

Exploring new methods like chemical vapor deposition using boron precursors or bottom-up approaches for scalable and controllable borophene synthesis. An innovative approach has been suggested. This technology closely resembles the electrochemical exfoliation process used to produce graphene. However, graphite exhibits conductivity at room temperature, but boron is nearly non-conductive at low temperatures. Hence, a novel design is being investigated, incorporating a heating coil with boron attachment, which serves as a conductive material that exhibits enhanced conductivity with rising temperatures. This design has been successfully employed to synthesize borophene from boron using an electrochemical exfoliation technique [39]. Another work reported the synthesis of borophene nanoparticles (NPs) from pure boron in three different types of liquids namely acetone, methanol, and benzyl using a sonochemical method. The synthesized borophene was further characterized by its physical, optical, and electronic behavior [82]. Few-layer borophene can be produced from bulk boron by the method-assisted liquid-phase exfoliation similar to that from Hummers-modified graphene synthesis [83]. Recent developments are being made in the liquid-phase exfoliation technique for synthesizing freestanding borophene nanosheets [84].

4.2 Enhancing Structural Stability and Electronic Properties

Doping with other elements or exploring heterostructures with other materials can improve stability and tailor electronic properties for specific applications. Predictions suggest that utilizing monovalent functional groups (F, Cl, Br, I, OH, and NH_2) to double the size of functionalization can effectively stabilize borophene, which has a structure similar to graphene and also allows for the manipulation of its electrical characteristics. Functionalized borophenes exhibit semiconductor properties and possess a diverse spectrum of band gaps. These gaps can be further adjusted by applying external pressures, making them very suitable for a variety of nanoelectronics applications [85]. In its free-standing condition, honeycomb borophene (hc-B) displays intrinsic instability due to partial occupation of both the in-plane and out-of-plane σ bands. The adsorption of light metals onto the monolayer can modify the electronic structure of hc-B by inducing charge transfer. The finding that charge transfer reduces the bonding σ and π bands below the Fermi level stabilizes hc-B in the case of beryllium-adorned hc-B. Using graphene to provide isoelectronic behavior maximizes the stability of hc-B, according to an analysis of the density of states, band structure, and charge transfer. The phonon dispersion of the beryllium-adorned structure not only confirms the relative stability as compared to free-standing hc-B but also offers additional proof [86]. Research has focused on solid interfaces of 2D materials and metal surfaces (m-surf). The borophene bilayer (BBL) is a promising material for 2D electrical devices. A study investigated the structural, electrical, and kinetic properties of the BBL/m-surf interface. This research utilizes Ag(111), Au(111), and Al(111) as m-surf surfaces. Additionally, the electronic transport characteristics of BBL channels coupled with BBL/m-surf top contacts were investigated. Metalization of BBL takes place via hybridization with metal surface states. Ohmic contacts develop between the BBL and the surface of the metal making it suitable for several electronic devices [87]. While the metallic property of borophene remains unaltered during the process of hydrogenation, the electronic structures undergo substantial modifications due to the placement and number of hydrogen atoms introduced into the borophene framework [88].

4.3 Exploring New Applications

Investigating the potential of borophene in areas like spintronics, photovoltaics, and water desalination, leveraging its unique properties for diverse technological advancements. The investigation of experimentally produced synthetic 2D-Xenes of the group-IIIA material (Borophene) that were predicted theoretically has been conducted. The 2D borophene Xene demonstrates metallic characteristics, making them capable of serving as contenders for forthcoming electrocatalytic nanodevices, specifically electrode materials, energy storage materials in batteries and superconductors, and catalysts [89]. Borophene is considered a highly auspicious and nascent substance with a wide array of potential applications, encompassing energy, sensors, and biological sectors. As an electrode material, borophene is a compelling alternative to graphene in high-performance supercapacitors due to its superior flexibility, capacitance, and conductivity. These characteristics have enabled the fabrication of high-performance supercapacitors for storage applications [90]. Borophene has high tensile strength, thermal and electrical conductivity, capacitance, and metallic properties. The remarkable properties of Borophene make it useful in thin-layer electronics, optoelectronics, capacitors, and biosensors [91]. In disease diagnosis and treatment strategies, borophene is a feasible substitute for graphene due to its comparable efficacy. Borophene is suitable for in vivo biodegradable applications due to its limited stability. Borophene exhibits significant potential as a nanomedicine for therapeutic applications, targeted drug delivery, the prevention of infectious diseases, and enhanced cancer therapy. It is imperative to surmount challenges before applying theoretical knowledge in clinical settings [92].

5 Upcoming Smart Research in Borophene

Particularly, the potential of 2D boron as a unique material and as a conceptual precursor for boron nanostructures of different dimensions makes it of special importance. Moreover, 2D boron is a good substitute for graphene, h-BN, and transition metal dichalcogenides, which are predicted to be essential building blocks in the next devices, because of their metallic characteristics. Still, borophenes are less stable and show intrinsic polymorphism than all bulk phases, which creates a production hurdle [46]. The future of borophene research hinges on "smart" approaches that combine theory, experiment, and computational modeling.

5.1 Integration of Borophene with Other Materials

Integrating borophene with complementary materials like graphene or transition metal dichalcogenides can create novel functionalities and synergistic effects for advanced devices. The Quantum Espresso software was used to analyze the potential of a new β_{12}-borophene/graphene (β_{12}-B/G) van der Waals (vdW) heterostructure. This study applied vdW-corrected density functional theory. Our investigation focuses on thermal and dynamical stability, adsorption energy, open circuit voltage, specific capacity, and diffusion barrier energy properties. The specific capacity reached an astounding 907 mAh/g, surpassing that of other 2D materials and heterostructures [93]. On Cu substrates, chemical vapor deposition epitaxially grows vertical borophene and graphene heterostructures above multilayer graphene for new device functionality. The carrier gas and boron sources are hydrogen and $NaBH_4$. The lattice structure of the borophene matches the predicted α-boron sheet. The borophene photodetector can detect UV (255 nm) and infrared (940 nm) light. In-situ integration of borophene with graphene is more sensitive than borophene or graphene photodetectors [94]. Researchers have developed a borophene-based electrochemoresistance (ECR) sensor that can enhance electrochemical sensor functionality and yield device benefits. The reported study offers a novel sensor design that uses the outstanding electrochemical activity and electrical characteristics of borophene. The presence of gases alters the electrical conductivity and electrochemical activity of borophene. The unique properties of boron allow for the creation of a new gas sensor [95]. The borophene surface demonstrates a substantial binding strength for target gas molecules, including CO, NO, NO_2, NH_3, and CO_2. Our estimated binding energies are greater than those of other 2D materials, including graphene, MoS_2, and phosphorene. The vibrational spectra of the gases are calculated when they are adsorbed on top of borophene to further analyze the sensitivity of borophene to these gases. Significant deviations from the reference values of the gases in their gas phase are evident in these spectra. The measurements of current–voltage (I-V) characteristics indicate the presence or absence of gas molecules, which function as ON and OFF states. This is compelling evidence that a gas detection device based on borophene is feasible [25] (Fig. 3).

The process of fabricating durable self-assembled film electrodes for zinc-ion capacitors (ZICs) using $Ti_3C_2T_x$-based MXene/B requires the eutectic etching of hydroxylated borophene, which serves as a trifunctional mediator. To promote efficient energy storage at the interface and an enhanced modification effect, the presence of this mediator facilitates a strong and cohesive bond between the various components of the heterostructure [96]. An electrophoretic deposition approach

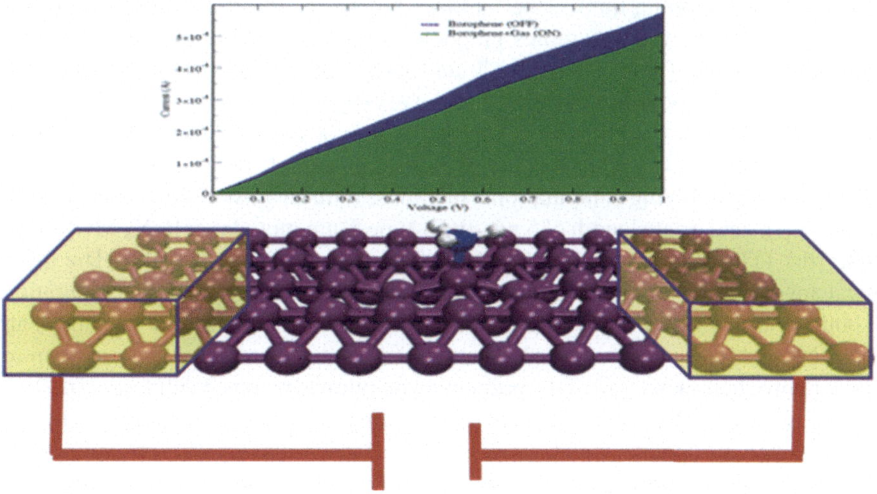

Fig. 3 The metallic properties of borophene enable the design of electrode configurations, resulting in significant changes in transmission functions compared to the original borophene surface. Reprinted (adapted) with permission from [25]. Copyright 2017 American Chemical Society

creates conductive 2D MXene/Borophene (MxB) electrodes for a flexible, high-speed supercapacitor. Electrostatic forces join positively charged Borophene and negatively charged MXene nanosheets. This arrangement keeps MXene nanosheets from stacking, increasing the layer gap. This assemblage exposes more electroactive regions, increasing electrolyte ion diffusion. With 85.14% capacitance retention at 20 $A{\cdot}g^{-1}$, the MxB (50:50) electrode, unsupported, shows remarkable performance. It has a long cycle life and a high gravimetric capacitance of 626.7 $F{\cdot}g^{-1}$ at 1 $A{\cdot}g^{-1}$. The developed supercapacitor has a maximum energy density of 75.6 $Wh{\cdot}kg^{-1}$ in comparison with carbon or MXene materials in aqueous electrolytes [97]. The fabrication of borophene/bismuthene/$Ti_3C_2T_x$ MXene/black phosphorus (BP) coatings involves the application of multiple layers via the spin-coating process. The result is the subsequent fabrication of a sensor that integrates photoelectrochemical and electrostatic principles. According to the results, applying a few layers of borophene and bismuthene can restore the surface imperfections on the Si/Ag substrate. Furthermore, by applying additional layers of 2D BP and TiO_2 derived from the identical material, it is possible to precisely restore any remaining flaws on the $Ti_3C_2T_x$ layer. [98].

5.2 Investigation of the Unique Properties of Borophene for Quantum Computing

Exploring the potential of borophene through its tunable bandgap and spin–orbit coupling for developing qubits, the building blocks of quantum computers. Halogenating borophene that resembles graphene produces a stable monolayer family consisting of B_4X_4 (where X = F, Cl, and Br). The highly anisotropic semiconducting electrical topologies of these monolayers suggest that they may find utility in ultra-scaled MOSFET applications. This research endeavors to analyze the quantum transport properties of B_4X_4 monolayers functioning as high-performance (HP) 5-nm MOSFETs by employing computational methods. With an on-state current exceeding 3000 $\mu A/\mu m$ and a gate length of 5 nm, the n-type 5-nm monolayer B_4X_4 MOSFETs satisfy the ITRS requirements for high-performance (HP) devices [99]. The theoretical investigation of the nonlinear optical properties involved the computation of both polarizability and hyperpolarizability. The observed values of super salt ($P_7Ba_2NO_3$) doped on borophene demonstrated a substantial rise in comparison to those of borophene across all of the previously mentioned results. With a range of 0.41 eV to 0.795 eV, the hardness values of all the complexes were comparatively low in comparison to borophene, which possesses a hardness value of 0.905 eV. Additional findings revealed that the excitation energies of each complex were comparatively lower than those of borophene, which possessed greater excitation energy [100]. In a particular research, TM@(α-B/BP) (TM = Sc–Ni) is a *3d* transition metal (TM) intercalated -borophene/black phosphorus (α-B/BP) heterostructure that has its structural, electrical, and magnetic properties investigated. Based on the findings, TM@(α-B/

BP) compounds demonstrate the capacity to manifest three distinct magnetic characteristics: nonmagnetic, ferromagnetic (FM), and antiferromagnetic (AFM). TM atom selection influences the particular magnetic behavior. Moreover, the majority of these systems exhibit substantial anisotropic magnetic energy. An AFM semiconductor, Ti@(α-B/BP) demonstrates a Néel temperature of 470 K, which is considerably higher than the average ambient temperature. In addition, it is possible to modify the electrical and magnetic properties of TM@(α-B/BP)s through manipulation of the TM intercalation concentration. The results of the study provide a pragmatic strategy for developing extremely prospective candidates for information storage and electronic devices [101]. Recently, a study examined XSi_2N_4 (X = Mo and W) heterostructures for structural, mechanical, transport, and optical properties. The electrical contacts of XSi_2N_4 with metals can serve as the foundation for nanoelectronic devices with ultralow Schottky barrier height. Among the investigated heterostructures, the interfaces between H and H-$MoSi_2N_4$/borophene exhibit the highest tunneling efficiency. At these interfaces, the barrier heights are 2.1 and 1.52 eV, whereas the barrier widths are 0.8 and 1.04 Å. Furthermore, the tunneling probabilities of 21.3% and 36.4% for these interfaces, respectively, indicate a significant degree of efficacy in the injection of carriers. The work highlighted the potential of constructing power-efficient Ohmic devices through the utilization of $MoSi_2N_4$/borophene contact [102]. Applications of 2D boron nanostructures is illustrated in (Fig. 4) [36].

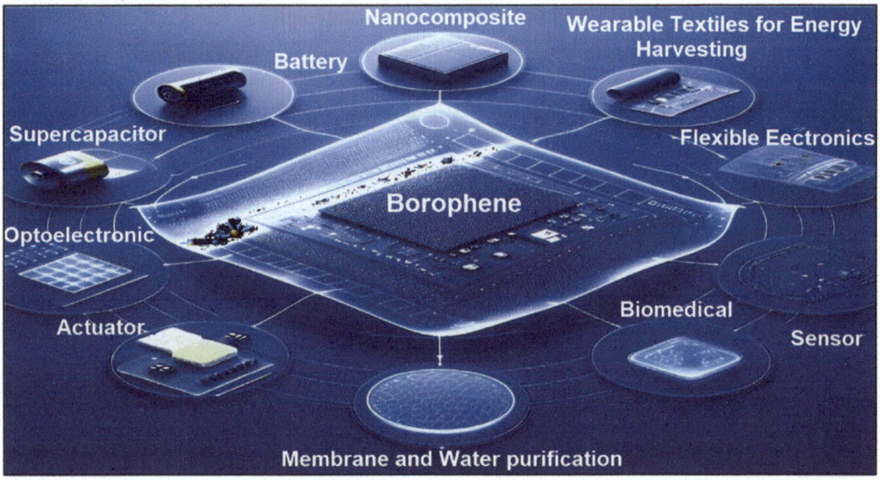

Fig. 4 2D boron nanostructures are transforming materials science and nanotechnology with a wide range of applications. Reprinted (adapted) with permission from [36]. Copyright 2024 Elsevier

5.3 Development of Borophene-Based Nanodevices for Biomedical Applications

Investigating the biocompatibility and unique properties of borophene-based sensors, biosensors, and detection of bio analytes and biomolecules in the surroundings. In a recent study, a borophene/zinc oxide (BZ) nanocomposite was fabricated using a simple solvothermal process. Hydrogels are created by integrating BZ nanocomposite into polymer networks through 3D micro extrusion printing and UV-induced free radical gelation. These hydrogels have compressibility, mechanical durability, pH sensitivity, and fissures. They undergo rigorous pH testing to ensure structural integrity. The hydrogels have remarkable elasticity, uniaxial mechanical properties, and non-toxic effects against Gram-positive and Gram-negative bacteria. They also show time-dependent and pH-dependent payload release, making them potential for use as drug carriers in biomedical research. This study is the first documented instance of a flexible biomaterial with antibacterial properties and efficient controlled drug delivery systems [103]. Moreover, another study demonstrated that borophene quantum dots (BPQDs) can be affixed to TiO_2 nanoparticles to enhance photoelectrochemical water-splitting activity. The process involves sonochemical production and hydrothermal affixation. The optimal concentration of BPQDs on TiO_2 nanoparticles is determined through UV–visible spectroscopy. The BPQDs, ranging from 2 to 9 nm in size, are characterized by boron structure. The nanocomposites show a decrease in photoluminescence emission intensity, suggesting a reduction in the time required for charge carriers to recombine. This results in enhanced electron liberation upon visible light exposure. The BPQDs/TiO_2 nanocomposite also improves photoelectrochemical water splitting due to the collective mechanisms of its components [104, 105]. This research uses borophene, a two-dimensional material, as a gas-sensing material for detecting NO_2 gas. The mixed matrix membrane (MMM), a selective gas separation membrane is formed using the Zeolitic imidazole framework (ZIF-8) and polyetherimide (PEI), which have specific solubility properties. The MMM's selectivity towards NO_2 gas is facilitated by its porosity and resistance to various compounds. The gas-sensing response of borophene remains negligible when exposed to 25 ppm NO_2 at 25 °C and 35% relative humidity, suggesting that the MMM filter doesn't significantly affect the sensory response. The hydrophobic properties of Borophene enhance stability under elevated humidity conditions. The study demonstrates that synthetically developed sensors can detect NO_2 gas at varying concentrations, crucial for ensuring human and environmental safety [105]. Furthermore, the synthesis of palladium on boron (Pd/B) was conducted in an ultrasonic water chamber, and its crystal structure and surface morphology were analyzed using SEM, HR-TEM, and XRD. The research found that Pd/B can reduce the cost of electrocatalysts used in ethanol oxidation, with a greater current and negative onset potential compared to Pd/C. The borophene in Pd/B enhances its tolerance and catalytic nature, resulting in a heightened capacity for electro-catalyzing EtOH, resulting in increased current density and extended stability [106].

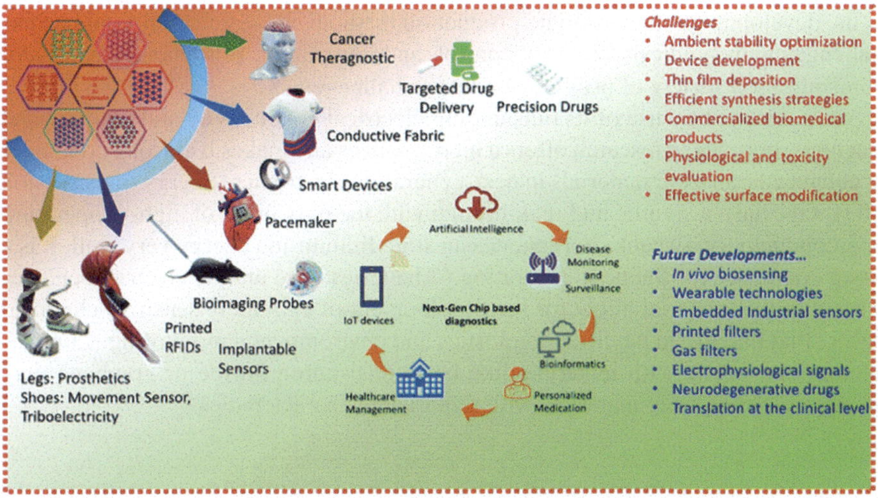

Fig. 5 Translational borophene in the 2D flatland has been achieved for biomedical applications. Reprinted (adapted) with permission from [92]. Copyright 2022 Royal Society of Chemistry

Borophene, a nanomaterial with significant biological properties, is being explored for its potential in biosensing, bioimaging, and cancer therapy. It is being used in advanced diagnostics by integrating AI and IoMT with high-performance platforms, allowing for personalized treatment and monitoring of human health. Borophene is also used in medical implants and prostheses due to its robustness, structural integrity, and biocompatibility. It has high potential as a nanomedicine for precise drug delivery, advanced cancer therapies, and infectious infections. However, researchers must overcome current hurdles before implementing theoretical skills in clinical practice. Borophene could be a powerful substitute for conventional 2D nanomaterials, meeting the pressing need for medications, in vivo biosensing devices, 3D printed structures, health monitoring gadgets, and high-performance energy storage devices as shown in Fig. 5 [92].

6 Future Challenges and Opportunities

Energy storage and electronics both seem to be promising applications for boron nanosheets. Still, complicated procedures and strict requirements make it difficult to produce excellent borophene sheets. Because of its intrinsic instability-especially when it comes into touch with moisture and oxygen-borophene cannot be used practically. Achieving particular features needs bandgap manipulation, and more research is needed to customize the energy bandgap. A further challenge is to transfer borophene sheets onto substrates while preserving their intrinsic characteristics. Complex devices utilizing borophene in combination with other materials are needed for the development of appropriate production methods. Research on borophene has challenges with synthesis, stability control, and integration. Borophene is still very promising for a variety of uses despite these challenges.

Alternatively, because of its outstanding electrical conductivity and unique band structures, borophene research offers a lot of chances and makes it a strong contender for enhanced transistor materials in next-generation electronics. Energy transmission and storage have a significant breakthrough with the possibility of high-temperature superconductivity. Because borophene can store lithium-ion energy very well, it is a strong contender for the next generation of batteries. The unique electrical characteristics of this material allow for the development of extremely sensitive chemical sensors that find use in medical diagnostics and environmental monitoring. Complementing borophene with materials like hexagonal boron nitride or graphene may provide new composite materials with remarkable characteristics.

6.1 Overcoming Technical and Practical Barriers in Borophene Research

Device integration: Integrating borophene with other materials and existing device architectures presents another hurdle. Researchers used time-domain nonadiabatic molecular dynamics simulations to study the photoexcited carrier dynamics of heterostructures made of β_{12}, χ_3, and α' borophene/MoS_2. Borophene-based heterostructures exhibit unique behavior due to Schottky interactions, electrical coupling, and phonon modes. The heterostructures have distinct characteristics for rapid carrier transfer. In α'/MoS_2 heterostructures, electrons tunnel into α' borophene at a rate of 29 femtoseconds. β_{12} borophene only allows hole migration for 176 femtoseconds, enabling efficient charge segregation. This study will be beneficial for 2D boron-based photovoltaic and optoelectronic devices [107]. The lattice mismatch between borophene and other materials can lead to defects and impede device performance. Developing strategies for the seamless integration of borophene with other components is crucial for realizing its potential in electronic and optoelectronic devices.

Cost-effectiveness: The current methods for synthesizing and processing borophene are often expensive and time-consuming. Scaling up production and reducing the cost will be essential for making borophene a viable material for commercial applications.

Environmental considerations: The synthesis and processing of borophene often involve hazardous chemicals and energy-intensive processes. Developing more sustainable and environmentally friendly approaches will be crucial for responsible development and implementation of borophene technology.

6.2 Collaborations and Interdisciplinary Approaches in Borophene Research

The research on borophene shows great potential, but substantial progress will necessitate a strong focus on collaboration and interdisciplinary methods. The following outlines how fostering collaboration can advance borophene research:

- **Collaboration between Material Scientists, Chemists, and Physicists**: Material scientists need to develop innovative synthesis methods for borophene production, collaborating with chemists to explore new precursors and reaction conditions. Chemistry is crucial for modifying borophene with specific groups for applications like sensors or catalysts. Physics emphasizes understanding fundamental electrical characteristics using theoretical calculations and modern characterization techniques. Collaborating with material scientists and chemists can guide in developing customized borophenes with customized functionalities.
- **Integration of Engineers and Nanotechnologists**: Engineers are developing procedures for transferring borophene sheets onto substrates and integrating them with other materials for device creation. Collaborating with material scientists ensures compatibility and preserves borophene features. Nanotechnologists manipulate nanoscale materials for intricate borophene-based structures, while physicists provide insights into their influence on electrical behavior.
- **Interdisciplinary Approaches**: Computational modeling aids in predicting borophene structures and directing experimental efforts toward specific functionality. Collaboration between physicists, chemists, and materials scientists is crucial for effective execution. Theoretical chemistry uses calculations to understand the chemical bonding and electrical properties of borophene, while experimental chemists verify predictions and enhance synthesis techniques.
- **Benefits of Collaboration**: Collaborating across various disciplines can expedite exploration and overcome obstacles in the field of borophene. This approach bridges knowledge gaps, allowing researchers to tackle its limitations from multiple perspectives. This exchange of knowledge can lead to innovative ideas and approaches.
- **Examples of Collaborative Research**: International research consortia and university-industry partnerships can unite scientists to address borophene-related problems, while university-industry partnerships can leverage academic expertise and industrial resources for borophene-based technology advancement.

6.3 Potential Commercialization and Industrial Applications of Borophene

Borophene, due to its distinctive characteristics, has significant potential for commercialization and utilization in several industrial areas. Nevertheless, substantial research and development efforts are required to overcome existing constraints and effectively transform its potential into tangible goods in practical settings. Here is a preview of several intriguing prospects:

- **Next-Generation Electronics**: The exceptional electrical conductivity and unique band structures of borophene make it a promising candidate for advanced transistors. These transistors could improve computer and mobile device processor speed and efficiency, reduce power consumption, and extend battery life. Additionally, their flexibility could lead to the development of flexible and wearable electronic systems.
- **Energy Storage and Conversion**: Borophene is a promising candidate for high-capacity advanced batteries due to its substantial theoretical lithium-ion storage capacity. These batteries could potentially extend the battery life of portable electronic devices and enhance the capacity and efficiency of grid-scale energy storage systems, particularly in renewable energy sources.
- **Sensors and Catalysis**: The unique electrical properties of borophene could be used to develop chemical sensors with high sensitivity, enhancing environmental monitoring and medical diagnostics. These sensors can identify pollutants in the atmosphere, water, and sediment, enhancing disease detection. Borophene can also be used as a catalyst for chemical reactions, enhancing their efficacy and sustainability in industry. This could lead to the development of innovative materials with advantageous properties.
- **Challenges and Considerations for Commercialization**: The commercial viability of borophene relies on developing cost-effective, scalable technologies, ensuring its stability, and finding ways to integrate it with other materials, with intellectual property playing a crucial role in its commercialization.

7 Conclusion

Borophene, a material with extraordinary characteristics, has significant potential for research due to its untapped potential, the need to bridge the gap between theory and reality, and the potential for synergistic advancements through a cooperative and multidisciplinary approach. Borophene can revolutionize quantum computing by enabling high-temperature superconductivity and topological characteristics, which are crucial for fault-tolerant quantum computers. It can also revolutionize biomedicine by using borophene-based drug delivery systems, biosensors, and biocompatible implants, improving disease diagnosis and treatment effectiveness. Furthermore, understanding and controlling its characteristics could lead to the development of next-generation materials with unique functions for various applications.

Borophene, a monolayer of boron atoms organized in a hexagonal lattice, has the potential to transform various industries due to its unique electronic characteristics, durability, and low mass. Its exceptional electrical conductivity could enable the development of faster and more efficient next-generation transistors, resulting in enhanced computational performance for computers and mobile devices. Its flexibility allows for the creation of flexible electrical devices, leading to innovative wearables and designs for consumer electronics. In energy storage and conversion, its theoretical capacity for storing high amounts of lithium-ion could lead to the creation of batteries with extended lifespans, transforming electric vehicles, handheld gadgets, and large-scale energy storage solutions. Additionally, borophene structures could demonstrate superconductivity at elevated temperatures, demonstrating significant advancements in energy transfer and storage. Its theoretical capacity to demonstrate high-temperature superconductivity could significantly impact quantum computing, as it can create fault-tolerant quantum bits (qubits) that exhibit less vulnerability to errors. This could lead to more reliable and capable quantum computer systems. Medical uses of borophene include drug delivery, biosensors, bioengineering, aviation, and automobile sectors. Its lightweight materials could enhance fuel efficiency and optimize performance in these industries. Its unique features could also be applied in water purification, chemical catalysis, and optoelectronics. Borophene research is expected to have a significant and widespread influence, causing a chain reaction of effects in various industries. With advancements in research and the resolution of existing obstacles in synthesis and integration, borophene holds the potential to revolutionize the technological domain and pave the way for a future filled with innovation and advancement.

The urgent appeal calls for increased funding for borophene research, fostering interdisciplinary collaboration among researchers, and bridging the gap between theoretical research and practical implementation, to fully unleash its potential. However, significant advancements are needed to turn theoretical promise into practical applications. Challenges include scalable synthesis, stability control, precise bandgap engineering, and integrating borophene with complementary materials. Scalable synthesis involves developing cost-effective methods for producing high-quality borophene sheets for practical applications. Stability control involves incorporating techniques like encapsulation or loading to improve device stability and operation. Precision bandgap engineering is crucial for customizing borophene for specific applications, such as transistors and optoelectronic devices. The integration of borophene with complementary materials is also crucial for operational device development. Prospective research domains include computational modeling, investigating novel borophene structures, strain engineering, and interface engineering. Advanced computational modeling methods can predict properties and guide experimental efforts, while strain engineering can reveal new functionalities and enhance borophene efficiency for specific applications. Understanding and manipulating the interface between borophene and other materials can optimize charge transfer and overall device performance.

Borophene research is a promising field that could revolutionize various industries by addressing challenges, identifying new avenues for research, and encouraging collaboration among various disciplines. This could lead to the development of high-capacity batteries for electric vehicles, highly sensitive biosensors for healthcare, lightweight materials for aerospace and automotive, and advanced water purification systems. By fostering strong partnerships between academic and industrial sectors, borophene research can lead to practical solutions.

Acknowledgements The authors express their sincere thanks to the Director, CSIR-AMPRI for his support and encouragement in this work. Shalu Yadav is thankful to CSIR, India for the award of SRF.

Conflict of Interest The authors declare no conflict of interest.

References

1. Li, L., et al.: Atomic-scale insights into the interlayer characteristics and oxygen reactivity of bilayer borophene. Angew. Chem. Int. Ed. **62**(32), e202306590 (2023)
2. Ayodhya, D., Veerabhadram, G.: A brief review on synthesis, properties and lithium-ion battery applications of borophene. FlatChem **19**, 100150 (2020). https://doi.org/10.1016/j.flatc.2019.100150
3. Mannix, A.J., Zhang, Z., Guisinger, N.P., Yakobson, B.I., Hersam, M.C.: Borophene as a prototype for synthetic 2D materials development. Nat. Nanotechnol. **13**(6), 444–450 (2018). https://doi.org/10.1038/s41565-018-0157-4
4. Hou, C., Tai, G., Wu, Z., Hao, J.: Borophene: current status, challenges and opportunities. ChemPlusChem **85**(9), 2186–2196 (2020). https://doi.org/10.1002/cplu.202000550
5. Liu, X., Li, Q., Ruan, Q., Rahn, M.S., Yakobson, B.I., Hersam, M.C.: Borophene synthesis beyond the single-atomic-layer limit. Nat. Mater. **21**(1), 35–40 (2022). https://doi.org/10.1038/s41563-021-01084-2
6. Zhang, Z., Yang, Y., Penev, E.S., Yakobson, B.I.: Elasticity, flexibility, and ideal strength of borophenes. Adv. Funct. Mater. **27**(9), 1605059 (2017). https://doi.org/10.1002/adfm.201605059
7. Yuan, J., Yu, N., Xue, K., Miao, X.: Ideal strength and elastic instability in single-layer 8-Pmmn borophene. RSC Adv. **7**(14), 8654–8660 (2017). https://doi.org/10.1039/C6RA28454J
8. Tran, T.-B.-T., Fang, T.-H., Nguyen, V.-T., Pham, V.-T.: Contact strength and deformation of straining free-standing borophene. Comput. Mater. Sci. **197**, 110624 (2021). https://doi.org/10.1016/j.commatsci.2021.110624
9. Gao, M., Li, Q.-Z., Yan, X.-W., Wang, J.: Prediction of phonon-mediated superconductivity in borophene. Phys. Rev. B **95**(2), 24505 (2017). https://doi.org/10.1103/PhysRevB.95.024505
10. Yan, L., et al.: Prediction of superconductivity in bilayer borophenes. RSC Adv. **11**(63), 40220–40227 (2021). https://doi.org/10.1039/D1RA08014H
11. Gao, Z., Li, M., Wang, J.-S.: Insight into two-dimensional borophene: five-center bond and phonon-mediated superconductivity. ACS Appl. Mater. Interfaces **11**(50), 47279–47288 (2019). https://doi.org/10.1021/acsami.9b17896
12. Zhang, Y., Yuan, X., Hao, J., Xu, M., Li, Y.: Realizing high-temperature superconductivity in borophene with Dirac states assembled by kagome and honeycomb boron layers. Mater. Today Phys. **35**, 101144 (2023). https://doi.org/10.1016/j.mtphys.2023.101144
13. Li, D., Chen, Y., He, J., Tang, Q., Zhong, C., Ding, G.: Review of thermal transport and electronic properties of borophene. Chin. Phys. B **27**(3), 36303 (2018). https://doi.org/10.1088/1674-1056/27/3/036303
14. Honari, N., Tabatabaei, S.M., Pourfath, M., Fathipour, M.: Semiconducting phase and anisotropic properties in borophene via chemical surface functionalization. J. Phys. Chem. C **124**(10), 5807–5816 (2020). https://doi.org/10.1021/acs.jpcc.9b06614

15. Zhong, H., Huang, K., Yu, G., Yuan, S.: Electronic and mechanical properties of few-layer borophene. Phys. Rev. B **98**(5), 54104 (2018). https://doi.org/10.1103/PhysRevB.98.054104
16. Lherbier, A., Botello-Méndez, A.R., Charlier, J.-C.: Electronic and optical properties of pristine and oxidized borophene. 2d Materials **3**(4), 45006 (2016). https://doi.org/10.1088/2053-1583/3/4/045006
17. Yar, A., Ilyas, A.: Effects of electric and magnetic fields on zitterbewegung of electron wave packet in borophene. J. Phys. Soc. Jpn. **89**(12), 124705 (2020). https://doi.org/10.7566/JPSJ.89.124705
18. Cruz, E., Carrillo-Bastos, R., Villavicencio, J.: Time diffraction and Zitterbewegung of two-dimensional massless Dirac excitations. Phys. Rev. B **101**(11), 115415 (2020). https://doi.org/10.1103/PhysRevB.101.115415
19. Liu, B., Zhou, K.: Recent progress on graphene-analogous 2D nanomaterials: properties, modeling and applications. Prog. Mater. Sci. **100**, 99–169 (2019). https://doi.org/10.1016/j.pmatsci.2018.09.004
20. Agarwal, A., Vitiello, M.S., Viti, L., Cupolillo, A., Politano, A.: Plasmonics with two-dimensional semiconductors: from basic research to technological applications. Nanoscale **10**(19), 8938–8946 (2018). https://doi.org/10.1039/C8NR01395K
21. Wang, Z., Lü, T.-Y., Wang, H.-Q., Feng, Y.P., Zheng, J.-C.: High anisotropy of fully hydrogenated borophene. Phys. Chem. Chem. Phys. **18**(46), 31424–31430 (2016). https://doi.org/10.1039/C6CP06164H
22. Faghihnasiri, M., Jafari, H., Ramazani, A., Shabani, M., Estalaki, S.M., Larson, R.G.: Nonlinear elastic behavior and anisotropic electronic properties of two-dimensional borophene. J. Appl. Phys. **125**(14), 145107 (2019). https://doi.org/10.1063/1.5079932
23. Mortazavi, B., Silani, M., Podryabinkin, E.V., Rabczuk, T., Zhuang, X., Shapeev, A.V.: First-principles multiscale modeling of mechanical properties in graphene/borophene heterostructures empowered by machine-learning interatomic potentials. Adv. Mater. **33**(35), 2102807 (2021). https://doi.org/10.1002/adma.202102807
24. Wang, H., Li, Q., Gao, Y., Miao, F., Zhou, X.-F., Wan, X.G.: Strain effects on borophene: ideal strength, negative Possion's ratio and phonon instability. New J. Phys. **18**(7), 73016 (2016)
25. Shukla, V., Wärnå, J., Jena, N.K., Grigoriev, A., Ahuja, R.: Toward the realization of 2D borophene based gas sensor. J. Phys. Chem. C **121**(48), 26869–26876 (2017). https://doi.org/10.1021/acs.jpcc.7b09552
26. Ranjan, P., et al.: Freestanding borophene and its hybrids. Adv. Mater. **31**(27), 1900353 (2019). https://doi.org/10.1002/adma.201900353
27. Alvarez-Quiceno, J.C., Miwa, R.H., Dalpian, G.M., Fazzio, A.: Oxidation of free-standing and supported borophene. 2d Materials **4**(2), 25025 (2017). https://doi.org/10.1088/2053-1583/aa55b6
28. He, J., Hu, Y., Li, D., Chen, J.: Ultra-low lattice thermal conductivity and promising thermoelectric figure of merit in borophene via chlorination. Nano Res. **15**(4), 3804–3811 (2022). https://doi.org/10.1007/s12274-021-3908-8
29. Yin, Y., et al.: Abnormal thermal conductivity enhancement in covalently bonded bilayer borophene allotrope. Nano Res. **15**(4), 3818–3824 (2022). https://doi.org/10.1007/s12274-021-3921-y
30. Hou, C., Tai, G., Liu, Y., Wu, Z., Liang, X., Liu, X.: Borophene-based materials for energy, sensors and information storage applications. Nano Res. Energy **2**, e9120051 (2023). https://doi.org/10.26599/NRE.2023.9120051
31. Li, C., et al.: Highly efficient, remarkable sensor activity and energy storage properties of MXenes and borophene nanomaterials. Prog. Solid State Chem. **70**, 100392 (2023). https://doi.org/10.1016/j.progsolidstchem.2023.100392
32. Chand, H., Kumar, A., Krishnan, V.: Borophene and boron-based nanosheets: recent advances in synthesis strategies and applications in the field of environment and energy. Adv. Mater. Interfaces **8**(15), 2100045 (2021). https://doi.org/10.1002/admi.202100045
33. Qiao, H., et al.: Tunable electronic and optical properties of 2D monoelemental materials beyond graphene for promising applications. Energy Environ. Mater. **4**(4), 522–543 (2021). https://doi.org/10.1002/eem2.12154

34. Huang, Z., Qi, X., Zhong, J.: 2D Monoelemental Materials (Xenes) and Related Technologies: Beyond Graphene. CRC Press (2022)
35. Ou, M., et al.: The emergence and evolution of borophene. Adv. Sci. **8**(12), 2001801 (2021). https://doi.org/10.1002/advs.202001801
36. Mishra, R.K., Sarkar, J., Verma, K., Chianella, I., Goel, S., Nezhad, H.Y.: Borophene: A 2D wonder shaping the future of nanotechnology and materials science. Nano Mater. Sci. (2024). https://doi.org/10.1016/j.nanoms.2024.03.007
37. Abdi, Y., et al.: A two-dimensional borophene supercapacitor. ACS Mater. Lett. **4**(10), 1929–1936 (2022). https://doi.org/10.1021/acsmaterialslett.2c00475
38. Srivastava, A., Dkhar, D.S., Singh, N., Azad, U.P., Chandra, P.: Exploring the potential applications of engineered borophene in nanobiosensing and theranostics. Biosensors **13**(7) (2023). https://doi.org/10.3390/bios13070740
39. Chowdhury, M.A., Uddin, M.M.K., Shuvho, M.B.A., Rana, M., Hossain, N.: A novel temperature dependent method for borophene synthesis. Appl. Surf. Sci. Adv. **11**, 100308 (2022)
40. Kiraly, B., et al.: Borophene synthesis on Au(111). ACS Nano **13**(4), 3816–3822 (2019). https://doi.org/10.1021/acsnano.8b09339
41. Guo, Z., et al.: Low-pressure CVD synthesis of tetragonal borophene single-crystalline sheets with high ambient stability. Cryst. Growth Des. **23**(6), 4506–4513 (2023). https://doi.org/10.1021/acs.cgd.3c00265
42. Ruan, Q., Wang, L., Bets, K.V., Yakobson, B.I.: Step-edge epitaxy for borophene growth on insulators. ACS Nano **15**(11), 18347–18353 (2021). https://doi.org/10.1021/acsnano.1c07589
43. Li, W., Wu, K., Chen, L.: Epitaxial growth of borophene on substrates. Prog. Surf. Sci. **98**(2), 100704 (2023). https://doi.org/10.1016/j.progsurf.2023.100704
44. Li, W., et al.: Experimental realization of honeycomb borophene. Sci. Bull. (Beijing) **63**(5), 282–286 (2018). https://doi.org/10.1016/j.scib.2018.02.006
45. Xu, R., Zou, X., Liu, B., Cheng, H.-M.: Computational design and property predictions for two-dimensional nanostructures. Mater. Today **21**(4), 391–418 (2018). https://doi.org/10.1016/j.mattod.2018.03.003
46. Zhang, Z., Penev, E.S., Yakobson, B.I.: Two-dimensional boron: structures, properties and applications. Chem. Soc. Rev. **46**(22), 6746–6763 (2017). https://doi.org/10.1039/C7CS00261K
47. Mannix, A.J., et al.: Synthesis of borophenes: anisotropic, two-dimensional boron polymorphs. Science (1979) **350**(6267), 1513–1516 (2015). https://doi.org/10.1126/science.aad1080
48. Li, D., et al.: 2D boron sheets: structure, growth, and electronic and thermal transport properties. Adv. Funct. Mater. **30**(8), 1904349 (2020). https://doi.org/10.1002/adfm.201904349
49. Rubab, A., Baig, N., Sher, M., Sohail, M.: Advances in ultrathin borophene materials. Chem. Eng. J. **401**, 126109 (2020). https://doi.org/10.1016/j.cej.2020.126109
50. Bhavyashree, M., Rondiya, S.R., Hareesh, K.: Exploring the emerging applications of the advanced 2-dimensional material borophene with its unique properties. RSC Adv. **12**(19), 12166–12192 (2022). https://doi.org/10.1039/D2RA00677D
51. Leonard Deepak, F., Casillas, G.: Recent highlights in advanced transmission electron microscopy techniques: applications to nanomaterials. In: Thomas, P.J., Revaprasadu, N. (eds.) Nanoscience, vol. 4. The Royal Society of Chemistry (2017). https://doi.org/10.1039/9781782620358-00029
52. Mendes, R.G., et al.: Electron-driven in situ transmission electron microscopy of 2D transition metal dichalcogenides and their 2D heterostructures. ACS Nano **13**(2), 978–995 (2019). https://doi.org/10.1021/acsnano.8b08079
53. Ranjan, P., et al.: 2D materials: increscent quantum flatland with immense potential for applications. Nano Converg **9**(1), 26 (2022). https://doi.org/10.1186/s40580-022-00317-7

54. Ashraf, N., Abghoui, Y.: Borophene potential for developing next-generation battery applications: a comprehensive review. Energy Fuels **37**(19), 14589–14603 (2023). https://doi.org/10.1021/acs.energyfuels.3c02319
55. Li, C., et al.: Two dimensional borophene nanomaterials: recent developments for novel renewable energy storage applications. Prog. Solid State Chem. **71**, 100416 (2023). https://doi.org/10.1016/j.progsolidstchem.2023.100416
56. Chaudhary, V., Kaushik, A., Furukawa, H., Khosla, A.: Review—towards 5th generation AI and IoT driven sustainable intelligent sensors based on 2D MXenes and borophene. ECS Sens. Plus **1**(1), 13601 (2022). https://doi.org/10.1149/2754-2726/ac5ac6
57. Rahman, A., et al.: Emerging 2D borophene: synthesis, characterization, and sensing applications. Sens. Actuators Phys. **359**, 114468 (2023). https://doi.org/10.1016/j.sna.2023.114468
58. Joseph, J., et al.: Borophene and boron fullerene materials in hydrogen storage: opportunities and challenges. Chemsuschem **13**(15), 3754–3765 (2020). https://doi.org/10.1002/cssc.202000782
59. Kong, L., Wu, K., Chen, L.: Recent progress on borophene: growth and structures. Front Phys. (Beijing) **13**(3), 138105 (2018). https://doi.org/10.1007/s11467-018-0752-8
60. Cuxart, M.G., et al.: Borophenes made easy. Sci. Adv. **7**(45), eabk1490 (2024). https://doi.org/10.1126/sciadv.abk1490
61. Chen, C., et al.: Synthesis of bilayer borophene. Nat. Chem. **14**(1), 25–31 (2022). https://doi.org/10.1038/s41557-021-00813-z
62. Ayub, M.N., et al.: Modern innovations in the provision and efficient application of 2D inorganic nanoscale materials (2024). https://doi.org/10.1515/revic-2023-0036
63. Patel, R.B.: Synthesis and Characterization of Novel Boron-Based Nanostructures and Composites. New Jersey Institute of Technology (2013)
64. Zhong, Q., et al.: Synthesis of borophene nanoribbons on Ag(110) surface. Phys. Rev. Mater. **1**(2), 21001 (2017). https://doi.org/10.1103/PhysRevMaterials.1.021001
65. Li, Q., Liu, X., Aklile, E.B., Li, S., Hersam, M.C.: Self-assembled borophene/graphene nanoribbon mixed-dimensional heterostructures. Nano Lett. **21**(9), 4029–4035 (2021). https://doi.org/10.1021/acs.nanolett.1c00909
66. Xu, Y., Xuan, X., Yang, T., Zhang, Z., Li, S.-D., Guo, W.: Quasi-freestanding bilayer borophene on Ag(111). Nano Lett. **22**(8), 3488–3494 (2022). https://doi.org/10.1021/acs.nanolett.1c05022
67. Vinogradov, N.A., Lyalin, A., Taketsugu, T., Vinogradov, A.S., Preobrajenski, A.: Single-phase borophene on Ir(111): formation, structure, and decoupling from the support. ACS Nano **13**(12), 14511–14518 (2019). https://doi.org/10.1021/acsnano.9b08296
68. Mozvashi, S.M., Givi, M.R., Tagani, M.B.: The effects of substrate and stacking in bilayer borophene. Sci. Rep. **12**(1), 13661 (2022). https://doi.org/10.1038/s41598-022-18076-0
69. Peng, B., et al.: Stability and strength of atomically thin borophene from first principles calculations. Mater. Res. Lett. **5**(6), 399–407 (2017)
70. Joshi, D.J., Malek, N.I., Kailasa, S.K.: Borophene as a rising star in materials chemistry: synthesis, properties and applications in analytical science and energy devices. New J. Chem. **46**(10), 4514–4533 (2022). https://doi.org/10.1039/D1NJ05271C
71. Silvestre, G.H., Scopel, W.L., Miwa, R.H.: Electronic stripes and transport properties in borophene heterostructures. Nanoscale **11**(38), 17894–17903 (2019). https://doi.org/10.1039/C9NR05279H
72. Gao, N., Wu, X., Jiang, X., Bai, Y., Zhao, J.: Structure and stability of bilayer borophene: the roles of hexagonal holes and interlayer bonding. FlatChem **7**, 48–54 (2018). https://doi.org/10.1016/j.flatc.2017.08.008
73. Ranjan, P., Lee, J.M., Kumar, P., Vinu, A.: Borophene: new sensation in flatland. Adv. Mater. **32**(34), 2000531 (2020). https://doi.org/10.1002/adma.202000531
74. Xu, L.-C., Du, A., Kou, L.: Hydrogenated borophene as a stable two-dimensional Dirac material with an ultrahigh Fermi velocity. Phys. Chem. Chem. Phys. **18**(39), 27284–27289 (2016). https://doi.org/10.1039/C6CP05405F

75. Gozar, A., Bozovic, I., Wu, R.: Quo vadis, borophene? Nano Today **50**, 101856 (2023). https://doi.org/10.1016/j.nantod.2023.101856
76. Xie, Z., et al.: Two-dimensional borophene: properties, fabrication, and promising applications. Research (2020). https://doi.org/10.34133/2020/2624617
77. Li, L., Schultz, J.F., Mahapatra, S., Lu, Z., Zhang, X., Jiang, N.: Chemically identifying single adatoms with single-bond sensitivity during oxidation reactions of borophene. Nat. Commun. **13**(1), 1796 (2022). https://doi.org/10.1038/s41467-022-29445-8
78. Bin Mujib, S., Ren, Z., Mukherjee, S., Soares, D.M., Singh, G.: Design, characterization, and application of elemental 2D materials for electrochemical energy storage, sensing, and catalysis. Mater. Adv. **1**(8), 2562–2591 (2020). https://doi.org/10.1039/D0MA00428F
79. Hu, P.-J., Ding, J.-T., Liang, Z.-R., Fang, T.-F., Guo, A.-M., Sun, Q.-F.: Enhanced electron transport and self-similarity in quasiperiodic borophene nanoribbons with line defects. Nanoscale **15**(25), 10740–10748 (2023). https://doi.org/10.1039/D3NR01658G
80. Anirudh, S., Krishnamurthy, S., Kandasubramanian, B., Kumar, P.: Probing into atomically thin layered nano-materials protective coating for aerospace and strategic defence application—a review. J. Alloys Compd. **968**, 172203 (2023). https://doi.org/10.1016/j.jallcom.2023.172203
81. Hu, P.-J., et al.: Resonant tunneling in disordered borophene nanoribbons with line defects. NPJ Comput. Mater. **8**(1), 131 (2022). https://doi.org/10.1038/s41524-022-00816-5
82. Asaduzzaman Chowdhury, M., et al.: Synthesis and characterization of borophene for future applications. Results Chem. **6**, 101071 (2023). https://doi.org/10.1016/j.rechem.2023.101071
83. Bingöl, M.S., Yilmaz, M., Kasapoğlu, A.E.: "Hummers' method-assisted liquid-phase exfoliation enables the fabrication of few-layer borophene from bulk boron. Turk J. Chem. **48**(2), 289–298 (2024)
84. Borah, P., Baruah, D.J., Mridha, P., Baishya, R., Bora, H.K., Das, M.R.: Photoenhanced intrinsic peroxidase-like activity of a metal-free biocompatible borophene photonanozyme for colorimetric sensor assay of dopamine biomolecule. Chem. Commun. **60**(17), 2417–2420 (2024). https://doi.org/10.1039/D3CC06326G
85. Tang, X., Gu, J., Shang, J., Chen, Z., Kou, L.: Double-sided surface functionalization: an effective approach to stabilize and modulate the electronic structure of graphene-like borophene. InfoMat **3**(3), 327–336 (2021). https://doi.org/10.1002/inf2.12126
86. John, D., Nharangatt, B., Chatanathodi, R.: Stabilizing honeycomb borophene by metal decoration: a computational study. J. Mater. Chem. C Mater. **7**(37), 11493–11499 (2019). https://doi.org/10.1039/C9TC03628H
87. Scopel, W.L., Crasto de Lima, F., Souza, P.H., Padilha, J.E., Miwa, R.H.: Bridging borophene and metal surfaces: structural, electronic, and electron transport properties. J. Phys. Chem. C **127**(35), 17556–17566 (2023). https://doi.org/10.1021/acs.jpcc.3c03123
88. Pang, K., et al.: First-principles calculations for the impact of hydrogenation on the electron behavior and stability of borophene nanosheets: implications for boron 2D electronics. ACS Appl Nano Mater. **5**(1), 1419–1425 (2022). https://doi.org/10.1021/acsanm.1c03956
89. Khan, K., et al.: Synthesis, properties and novel electrocatalytic applications of the 2D-borophene Xenes. Prog. Solid State Chem. **59**, 100283 (2020). https://doi.org/10.1016/j.progsolidstchem.2020.100283
90. Sahoo, B.B., Pandey, V.S., Dogonchi, A.S., Thatoi, D.N., Nayak, N., Nayak, M.K.: Exploring the potential of borophene-based materials for improving energy storage in supercapacitors. Inorg. Chem. Commun. **154**, 110919 (2023). https://doi.org/10.1016/j.inoche.2023.110919
91. Gupta, G.H., Kadakia, S., Agiwal, D., Keshari, T., Kumar, S.: Borophene nanomaterials: synthesis and applications in biosensors. Mater. Adv. **5**(5), 1803–1816 (2024). https://doi.org/10.1039/D3MA00829K
92. Yadav, S., Sadique, M.A., Kaushik, A., Ranjan, P., Khan, R., Srivastava, A.K.: Borophene as an emerging 2D flatland for biomedical applications: current challenges and future prospects. J. Mater. Chem. B **10**(8) (2022). https://doi.org/10.1039/d1tb02277f
93. Faramarzi, S., Movlarooy, T.: β12-borophene/graphene heterostructure as a high-performance anode material for Li-ion batteries. ACS Appl. Mater. Interfaces **16**(20), 25966–25976 (2024). https://doi.org/10.1021/acsami.3c17997

94. Wu, Z., et al.: Epitaxial growth of borophene on graphene surface towards efficient and broadband photodetector. Nano Res. **17**(4), 3053–3060 (2024). https://doi.org/10.1007/s12274-023-6109-9
95. Hosseini, S.M., et al.: Electrochemoresistance sensor: a borophene-based sensor with simultaneous electrochemical and chemoresistance sensing capability. ACS Mater. Lett. **6**(3), 933–942 (2024). https://doi.org/10.1021/acsmaterialslett.3c01120
96. Yong, B., et al.: A trifunctional hydroxylated borophene-mediated MXene enabled superstable and fast-kinetics interface storage. Adv. Funct. Mater. 2316127 (2024) https://doi.org/10.1002/adfm.202316127
97. Somesh, T.E., et al.: Flexible 2D borophene-stacked MXene heterostructure for high-performance supercapacitors. Chem. Eng. J. **481**, 148266 (2024). https://doi.org/10.1016/j.cej.2023.148266
98. Zeng, W., et al.: Defect engineering borophene/bismuthene/MXene/black phosphorus van der Waals laminations for sensitive photoelectrochemical-electrostatic sensors. Appl. Surf. Sci. **660**, 160009 (2024). https://doi.org/10.1016/j.apsusc.2024.160009
99. Guo, S., et al.: Theoretical dissection of the electronic anisotropy and quantum transport of ultrascaled halogenated borophene MOSFETs. Phys. Rev. Appl. **21**(5), 54016 (2024). https://doi.org/10.1103/PhysRevApplied.21.054016
100. Ayub, A.R., et al.: An in-depth investigation of nonlinear optical properties of self-assembled borophene doped with P7Ba2NO3: a computational simulating study. Phys. Scr. **99**(5), 55949 (2024). https://doi.org/10.1088/1402-4896/ad398a
101. Zhang, H., et al.: Structure, electronic, and magnetic properties of 3d transition metal intercalated borophene/BP heterostructures. Int. J. Quantum Chem. **124**(1), e27300 (2024). https://doi.org/10.1002/qua.27300
102. Jalil, A., Zhao, T., Firdous, A., Kanwal, A., Ali Raza, S.R., Rafiq, A.: Computational Insights into Schottky Barrier heights: graphene and borophene interfaces with H- and Ḣ-XSi2N4 (X = Mo, W) monolayers. Langmuir **40**(16), 8463–8473 (2024). https://doi.org/10.1021/acs.langmuir.3c04045
103. Das, P., et al.: Borophene based 3D extrusion printed nanocomposite hydrogel for antibacterial and controlled release application. Adv. Funct. Mater. **34**(21), 2314520 (2024). https://doi.org/10.1002/adfm.202314520
104. Yashwanth, H.J., Hareesh, K., Rondiya, S.R., Choudhary, R.J., Dhole, S.D.: The borophene quantum dots scaffolded TiO2 nanocomposite as an efficient photo electrocatalyst for water splitting application. Appl. Surf. Sci. **646**, 158910 (2024). https://doi.org/10.1016/j.apsusc.2023.158910
105. Arkoti, N.K., Pal, K.: Improved selectivity of borophene sensor towards NO2 gas with PEI-ZIF-8 overlayer. Sens. Actuators B Chem. **401**, 135033 (2024). https://doi.org/10.1016/j.snb.2023.135033
106. Mabhulusa, W., Sekhosana, K.E., Fuku, X.: Incorporation of Pd catalyst into highly effective borophene nanosheet Co-catalyst for electrokinetics and electrochemical oxygen reduction reactions. J. Electron. Mater. (2024). https://doi.org/10.1007/s11664-024-11113-w
107. Kang, Y., et al.: Selective interfacial excited-state carrier dynamics and efficient charge separation in borophene-based heterostructures. Adv. Mater. **36**(5), 2307591 (2024). https://doi.org/10.1002/adma.202307591

SPRINGER NATURE

GPSR Compliance

The European Union's (EU) General Product Safety Regulation (GPSR) is a set of rules that requires consumer products to be safe and our obligations to ensure this.

If you have any concerns about our products, you can contact us on ProductSafety@springernature.com

In case Publisher is established outside the EU, the EU authorized representative is:

Springer Nature Customer Service Center GmbH
Europaplatz 3
69115 Heidelberg, Germany

The manufacturer's authorised representative in the EU is Springer
Nature Customer Service Centre GmbH, Europaplatz 3, 69115 Heidelberg,
Germany. If you have any concerns regarding our products, please
contact ProductSafety@springernature.com

Printed and bound by CPI Group (UK) Ltd, Croydon, CR0 4YY
01/10/2026
02225906-0001